T0123012

Single Stage to Orbit

New Series in NASA History

Roger D. Launius, Series Editor

Related Books in the Series

NASA and the Space Industry
Joan Lisa Bromberg

Space Policy in the Twenty-First Century
edited by W. Henry Lambright

The Space Station Decision: Incremental Politics and Technological Choice
Howard E. McCurdy

Faster, Better, Cheaper: Low-Cost Innovation in the U.S. Space Program
Howard E. McCurdy

Single Stage to Orbit

Politics, Space Technology, and the Quest for Reusable Rocketry

Andrew J. Butrica

The Johns Hopkins University Press
Baltimore and London

Printed in the United States of America on acid-free paper
9 8 7 6 5 4 3 2 1

The Johns Hopkins University Press
2715 North Charles Street
Baltimore, Maryland 21218-4363
www.press.jhu.edu

Library of Congress Cataloging-in-Publication Data

Butrica, Andrew J.
Single stage to orbit : politics, space technology, and the quest for reusable rocketry / Andrew J. Butrica.
p. cm.— (New series in NASA history)
Includes bibliographical references and index.
ISBN 0-8018-7338-X (hardcover : alk. paper)
1. Aerospace planes—Research—United States. 2. Reusable space vehicles—Research—United States. 3. Hypersonic aerodynamics—Research—Political aspects—United States. 4. Research aircraft—United States. 5. United States. National Aeronautics and Space Administration. I. Title. II. Series.
TL795.5, B88 2003
629.44'1'0973-dc21 2002152156

A catalog record for this book is available from the British Library.

To George Peter Butrica (1913-1956),
U.S. Army Technician Fifth Grade, survivor of D-Day, Bronze Star recipient for "meritorious service in support of operations against the enemy in France," member of the United Brotherhood of Carpenters and Joiners of America, Local Union 1812, and father. He never lived to see the dawn of the space age.

"We set sail on this new sea because there is new knowledge to be gained, and new rights to be won, and they must be won and used for the progress of all people. For space science, like nuclear science and all technology, has no conscience of its own. Whether it will become a force for good or ill depends on man, and only if the United States occupies a position of pre-eminence can we help decide whether this new ocean will be a sea of peace or a new terrifying theater of war."
President John F. Kennedy, September 12, 1962

"To use a Southern euphemism, our space program has been snake-bit."
Sen. Albert Gore, Jr., May 5, 1986

Contents

Acknowledgments

This book would not have come into existence if the National Aeronautics and Space Administration (NASA) had not embarked on the ill-fated *X-33* project and had not contracted to have the project's development documented from beginning to end. The *X-33* program's stated goal was to reduce the business and technical risks by the year 2000 so that private industry could build and operate a fully reusable single-stage-to-orbit launch vehicle for civilian, military, and commercial customers. The payoff for NASA would be a dramatic reduction in the cost of putting payloads in space.

NASA selected the Lockheed Martin Skunk Works to build the *X-33* following a competition among three industry designs that included Rockwell International and McDonnell Douglas. The Skunk Works concept used a lifting body shape coupled with aerospike rocket engines. The *X-33* prototype would simulate the ascent and reentry environments of the full-scale launcher, flying over thirteen times the speed of sound (Mach 13) at altitudes approaching 50 miles. It would have made as many as fifteen flights from the Haystack Butte launch site, located near Edwards Air Force Base, California, to landing sites in Utah and Montana, beginning in June 1999. However, the Skunk Works postponed those flights several times and never completed construction of the vehicle. By March 2001, neither NASA nor Lockheed Martin wanted to spend more money on the project, or on detailing its history, and Lockheed Martin began dismantling the craft in January 2002. In March 1997, NASA hired me to document the *X-33* program. Among the many deliverables that the contract required was a monograph "on aspects of the development of the system (short booklet of less than 100 pages)" and a book-length "formal history of the project." The monograph was due early in the contract, while the completed book manuscript constituted the final deliverable. My initial thought was to focus the monograph on the propulsion system, that is, the aerospike engines. The exercise would have been helpful in preparing a small portion of the *X-33* history. However, the more I researched the *X-33* and single-stage-to-orbit rockets, the more I recognized that another project had been a decisive forerunner of the *X-33:* the *Delta Clipper Experimental (DC-X)* built by McDonnell Douglas for the Strategic Defense Initiative Organization. The *DC-X* story would constitute a large section of the *X-33* book-length history. I realized that if the *DC-X,* rather than the aerospike engine, were the monograph's subject, I would lay much more groundwork for the future *X-33* history book.

After receiving approval for the monograph outline that chronicled the *DC-X*'s development from concept to flight tests at White Sands, New Mexico, I wrote a manuscript titled "From the X-Rocket to the Delta Clipper: The Making and Unmaking of a Spaceship." With encouragement from Roger Launius, Chief Historian, and Gary Payton, Deputy Associate Administrator for Space, both at NASA Headquarters, I expanded the monograph into a book-length manuscript. Meanwhile, the failure of the *X-33* program was becoming increasingly apparent. When NASA terminated the *X-33* program and the history project, the manuscript history of the *DC-X,* rather than that of the *X-33,* became the contract's final deliverable.

This book, then, researched and written entirely under contract with NASA Headquarters, would not have come into existence without NASA's desire to document the *X-33* program's development. I owe much gratitude and appreciation to Roger Launius and Gary Payton, who oversaw the contract. They provided secretarial, telephone, photocopying, and other supplies and services, as well as a professional environment in which to write. The professional and congenial support of Roger and the History Office staff was extremely helpful. A special word of recognition also goes to Gary Payton. A believer in single-stage-to-orbit transport and in the importance of the *X-33* program, he convinced NASA to fund the history project in the first place. He was most encouraging, even when the going got rough, and provided many insights in our informal talks. It was a pleasure to work with both Roger and Gary.

This book owes much to those who donated boxes and boxes of documents over the course of the *X-33* history project. Together they formed the bulk of the collection of primary documents that became the *X-33* archive which served as the documentary basis for this history. Without these donations, this history literally would have been impossible to write, or at the very least, would have been a second-rate narrative. Gertrude "Trudy" E. Bell contributed a wealth of information on the space movement, a subject that she has followed and studied over the years. Cindy Bruno, formerly of NASA Headquarters, and Matt Crouch, NASA Headquarters, furnished a multitude of materials on a range of policy, business, and technological topics.

When he left the agency, Gary Payton bequeathed a substantial quantity of valuable primary records on the *X-33*, *X-34*, and other NASA programs. These were all the more useful because of his authoritative position over those programs. However, far more relevant to the *DC-X* and the SDIO's SSTO program were the wealth of documents that Tim Kyger and Steve Hoeser contributed from their personal collections. The value of those documents derived from their sheer quantity and the pivotal roles that both Kyger and Hoeser played in the *DC-X* story. While serving on Rep. Dana Rohrabacher's staff, Kyger orchestrated congressional efforts to keep the program funded and the vehicle flying. Hoeser served as Lt. Gen. Daniel Graham's aide-de-camp and Maxwell Hunter's double, especially during the crucial period of promoting the project in Washington, and he documented events extensively and meticulously. Hoeser's and Kyger's continuing cooperation and support throughout the *X-33* history project and beyond merit special recognition.

Many others also contributed documents. I wish to thank in particular NASA staff members James P. Arrington (Langley), William Claybaugh (Headquarters), Charles Eldred (Langley), Delma Freeman, Jr. (Langley), Ann Gaudreaux (Langley), Dill Hunley (Dryden), Kenneth Iliff (Dryden), Dale Mackall (Dryden), Douglas O. Stanley (Langley), Theodore A. Talay (Langley), and Alan Wilhite (Langley). In addition, I want to express my gratitude to those in industry, academia, and government who furnished documents: Arnold Aldrich (Lockheed Martin), Robert Baumgartner (Skunk Works), Michael Epstein (AlliedSignal), William Gaubatz (Universal Space Lines), Michael Griffin (Orbital Sciences), the late Maxwell Hunter (aerospace consultant), W. D. "Woody" Kay (Northeastern University), Paul L. Klevatt (McDonnell Douglas, retired), James Muncy (government consultant), J. Scott Reaser (Skunk Works), Jess Sponable (U.S. Air Force), and David Urie (Skunk Works). Of these, I want to acknowledge especially the personal papers supplied by Max Hunter and Woody Kay's materials from the Ronald Reagan Presidential Library.

Additionally, I want to express my appreciation to certain individuals who were the contacts for institutional document sources: Brenda Parker, office of the Associate Administrator for Space, Department of Transportation; Donald Baucom, historian, Missile Defense Agency; the staff of the Space Studies

Institute's Space Business Archive; and Colin Fries, Steve Garber, Mark Kahn, and Jane Odom of the NASA Historical Reference Collection. Finally, a word of thanks to all of those who built and maintained, and to those who provided "content" to, the websites of the Marshall Space Flight Center, General Accounting Office, NASA Inspector General, and Lockheed Martin's VentureStar.

The many oral history interviews conducted for the *X-33* history project also constituted an efficacious resource for the writing of this book. For their cooperation I want to thank: Arnold Aldrich, Robert "Gene" Austin, Robert Baumgartner, Ivan Bekey, James Berry, Raymond Chase, William Claybaugh, Charles "Pete" Conrad, Dan Dumbacher, Charles Eldred, Michael Epstein, Delma Freeman, Jr., William Gaubatz, Michael Griffin, Michael Hampson, James Hengle, Andrew Hepler, Steve Hoeser, Max Hunter, Kenneth Iliff, Paul Klevatt, Tim Kyger, Joel Leventhal, Jack Levine, John Mansfield, T. K. Mattingly, James Muncy, Gary Payton, Gregory Reck, Jerry Rising, Ronald Schena, Curtis Shoffner, Jess Sponable, Doug Stanley, Phil Sumrall, Ted Talay, David Urie, and Alan Wilhite.

I am also grateful to the many individuals who freely gave their time to read all or portions of this book (and its several manuscript predecessors): Wayne Anderson, Ivan Bekey, William Claybaugh, Erik Conway, Charles Eldred, Larry Gambone, William Gaubatz, Andrew Hepler, Steve Hoeser, Max Hunter, Paul Klevatt, Tim Kyger, Roger Launius, James Muncy, Michael Osborne, Gary Payton, Jay Penn, Jerry Pournelle, Ronald Schena, John Sloan, Jess Sponable, Doug Stanley, David Whalen, and the anonymous reviewer for Johns Hopkins University Press. Thanks also go to those who helped with finding photographs: Don Baucom, Bill Gaubatz, Colin Fries, Daniel Graham, Jr., Steve Hoeser, Paul Klevatt, Tim Kyger, Pat Ladner, Jane Odom, Jerry Pournelle, Ron Schena, and Jess Sponable.

Despite the help of many in reading and refining this text, the errors and inaccuracies contained in this book are entirely those of the author. Finally, I want to thank Prof. Fred Suppe, who kindly arranged for my access to the University of Maryland's library system through the Committee on the History and Philosophy of Science (CHPS), and Joann "CB" Carpenter, my confidante, dance partner, and soul mate.

Chronology

1972	May 26	Antiballistic Missile Treaty (SALT I)
1979	June 18	SALT II
1981	January 20	President Reagan sworn in
	January 30	First meeting of the Citizens' Advisory Council on National Space Policy
	August 5	First commercial launcher, the *Percheron,* explodes during engine test.
1982	July 4	NSDD 42 marks start of a national policy on space commerce
	September 9	Launch of *Conestoga I,* first successful U.S. commercial launch vehicle
1983	March 23	President Reagan announces the Strategic Defense Initiative (SDI)
	November 18	Designation of the Department of Transportation (DoT) as lead agency to promote U.S. commercial launch industry
1985	August 6	Max Hunter writes about the *X-Rocket*
1986	January 28	*Challenger* accident
1988	March 15	Hunter drafts first description of the *SSX*
	December 5	Daniel Graham volunteers to arrange a briefing on the *SSX* with Vice President Quayle
	December 8	Mikhail Gorbachev announces domestic reforms and troop withdrawals
1989	February 15	Graham, Hunter, and Pournelle brief Quayle on the *SSX*
	March 29	DoT issues first commercial launch license
	July 19	Aerospace Corporation evaluates the *SSX* concept
	November	Germans dismantle Berlin Wall, symbolically ending the cold war

1990	January	Estonia, Latvia, and Lithuania demand independence from the Soviet Union
	March 26	SDIO publishes RFP for the Single Stage To Orbit (SSTO) program
	July 2	Industry proposals due for Phase I
	August 8	Iraq annexes Kuwait
	August 15	SDIO SSTO program Phase I contract awards
1991	January 29	President Bush announces GPALS
	February 22	SDIO asks NASA Langley to evaluate Phase I SSTO contractor designs
	March 3	The Persian Gulf War (Desert Storm) ends
	June 5	SDIO opens Phase II competition
	August 16	SDIO awards Phase II contract to McDonnell Douglas to build and fly its *Delta Clipper Experimental (DC-X)* rocket
	December	Format I cuts *DC-X* funding
	December 8	Russia, Belarus, and Ukraine establish the Commonwealth of Independent States (CIS)
1992	February	Pentagon reactivates Format I *DC-X* funding cut
	February	The SSTO program passes Initial Design Review
	March 29	Final rescinding of Format I
	April 1	Daniel Goldin becomes NASA administrator
	June 23	SDIO releases the SSTO program Environmental Assessment
	June 30	*DC-X* Final Design Review takes place
	July 2	Space Frontier Foundation calls for the resignation of NASA Langley's William Piland
	September 24	Congress directs NASA to conduct the Access to Space study
1993	January 7	NASA's Access to Space study begins
	January 20	President Clinton and Vice President Gore sworn into office
	April 3	Rollout of the *DC-X*
	May 13	The SDIO becomes the Ballistic Missile Defense Organization (BMDO); the SSTO program becomes the Single Stage Rocket Technology (SSRT) program

	July	NASA's Access to Space study endorses an experimental single-stage-to-orbit program
	August 18	*DC-X* first flight
	August 31	NASA briefing on the *X-2000*
	September	H.R. 2401 transfers the SSRT program to DARPA
	September 1	The BMDO announces its intention to release a draft Request For Proposals for the SX-2
	September 11	*DC-X* second flight
	September 30	*DC-X* third flight
	September 30	*DC-X* flights halted for lack of money
	October 1	The Terry Dawson meeting takes place
	December 31	Pentagon recision abolishes SSRT program
1994	January 31	Daniel Goldin releases NASA funds to *DC-X* project
	February 23	NASA Research Announcements to convert *DC-X* into the *DC-XA*
	March 3	NASA announces winners of *DC-XA* awards
	April	DARPA releases *DC-X* funding to the BMDO
	May	Moorman Report issued
	May	NASA designated the lead agency for reusable launch systems, Pentagon the lead agency for expendables
	June 20	*DC-X* fourth flight
	June 27	*DC-X* fifth flight
1995	May 16	*DC-X* sixth flight
	June 12	*DC-X* seventh flight
	July 7	*DC-X* final flight
1996	March 15	*DC-XA* rollout
	May 4	*DC-XA* engine tests begin
	May 18	*DC-XA* first flight
	June 7	*DC-XA* second flight; first as the *Clipper Graham*
	June 8	*DC-XA* third flight
	July 2	Gore announces *X-33* Phase II winner
	July 31	*DC-XA* final flight
	September 12	NASA *DC-XA* mishap report released
2000	August 10	*X-33* liquid hydrogen tank failure report
2001	January 20	President George W. Bush sworn in

	March	NASA ends funding for the *X-33* program
2002	January 4	The BMDO becomes the Missile Defense Agency
	January	Lockheed Martin begins dismantling the *X-33*
	June 13	1972 ABM Treaty rescinded

Single Stage to Orbit

Introduction

"In a village of La Mancha, whose name I have no desire to remember, there lived not long ago one of those gentlemen who keep a lance in the lance-rack, an old buckler, a lean hack, and a greyhound for coursing." So begins Miguel de Cervantes's (1547–1616) tale of the epic adventures of Don Quixote de la Mancha. Cervantes intended for the adventures (and misadventures) of Don Quixote to lampoon the craze for chivalrous literature among his contemporaries. Hence, the story's hero appears ludicrous as he mistakes a shaving basin for the Golden Helmet of Mambrino or battles thirty "monstrous giants." Even after discovering that he has just charged a windmill, Don Quixote persists in his illusion. His enemy, he believes, has turned the giants into windmills in order to rob him of the glory of vanquishing them.

Like Cervantes's knight errant, the main characters in the story of this book were convinced of something that the vast majority of their peers rejected as an impossible and unobtainable chimera. They believed in the possibility of building and flying a reusable single-stage-to-orbit rocket. In one form or another, this idea has been taken seriously by rocketeers since the 1920s and has been (and continues to be) considered, by knowledgeable experts, technologically unrealizable. For a single-stage-to-orbit vehicle to succeed, its mass at launch has to be over 90 percent propellant (what aerospace engineers call "mass fraction"). As a result, the remaining structure, including wings, landing gear, and other hardware that are used only for reentry and landing, has to be built out of incredibly strong lightweight material. To date, no reusable single-stage-to-orbit rocket has taken off or landed on this planet. Although the National Aeronautics and Space Administration (NASA) endeavored to develop a single-stage-to-orbit transport in its *X-33* program, with that project's cancellation in March 2001, an operational reusable single-stage-to-orbit rocket still remains in the future. However, we are closer to achieving that goal because of the flights of an earlier experimental vehicle known as the *DC-X (Delta Clipper Experimental),* built by McDonnell Douglas for the Strategic Defense Initiative Organization (SDIO). This is the story of how a few visionaries, true believers in reusable single-stage-to-orbit rockets, turned a technological chimera into a concrete government program and flight vehicle.

Unlike Don Quixote's delusions, the vision of these prophets originated not in fantasies but in the down-to-earth need to reduce the cost of putting payloads in orbit. They were pragmatic, yet they also believed that endless possibilities for utilizing space would materialize, once low-cost access to orbit became a reality. Single-stage-to-orbit spaceships have the potential to lower launch costs dramatically from thousands of dollars to hundreds of dollars per payload pound. Once launch costs came down, the final frontier, space, would open up for whatever possibilities the mind might conceive: settlement, mining, tourism, manufacturing, or exploration.

Nearly all the contemporaries of these visionaries dismissed the single-stage-to-orbit goal as technological fantasy, in spite of its practicality. Their perspective was much like that of Cervantes's Sancho Panza. They saw the shaving basin, not the Golden Helmet of Mambrino. These aerospace engineers and managers understood how technologically difficult the stage-and-a-half shuttle was. If a fully reusable two-stage-to-orbit shuttle were in the future, it was not in the immediate future, and, certainly, a single-stage-to-orbit spacecraft was so far in the future as to be impossible. Like most engineers inside and outside the aerospace sector, they believed that technological change took place incrementally, not in giant leaps, and an operational single-stage-to-orbit vehicle was too much of a leap. They agreed that a single-stage-to-orbit spacecraft would reduce operational costs, but so would a Star Trek transporter, which they viewed as being equally unrealistic. Just how imaginative could one be inside the military-industrial complex?

Critics argue even today that single-stage-to-orbit spacecraft will remain in the domain of science fiction and fantasy for a long time into the future. Indeed, NASA has rejected single-stage-to-orbit architectures for its Space Launch Initiative, the agency's latest attempt to design a shuttle replacement. As for science fiction writers, they have venerated single-stage-to-orbit transport for nearly a century and a half, since at least Jules Verne's *From the Earth to the Moon,* which first appeared in 1865. It is not coincidental that science fiction writers played a role in the development of the single-stage-to-orbit rocket, for the single-stage-to-orbit story is one of the imagination and, therefore, one wrought with irony, the very heart of fiction. Single-stage-to-orbit spaceships are both the stuff of our imagination (because they do not yet exist) and the stuff that inspires our imagination. Our pragmatic visionaries justified their quest by the need to reduce the high cost of launching into space. Their vision lay in a "twilight zone" at once in the "dimension of the imagination" and in a reality-centered concern for launch economics. Once realized, the vision of single-stage-to-orbit transport would turn space visions into space realities and, in turn, have a profound impact on life on this planet.

The middle ground between the Quixotes and the Sancho Panzas, like the middle ground between technology and the imagination, lies in the realm of the attainable. The pragmatic dreamers could advance their cause by convincing others, especially corporate and government managers, that the technology for a single-stage-to-orbit vehicle was within reach. In this way, a small classified program or internal corporate study might continue the quest for a reusable single-stage-to-orbit spacecraft. Thus, despite overwhelming skepticism about the feasibility of single-stage-to-orbit transport, research and development dollars underwrote the vision.

This is not the story of one idea, single-stage-to-orbit transport, but of a vision that brought together several concepts, the first and oldest of which was the reusable, rocket-powered, single-stage-to-orbit vehicle. A second key idea was called "aircraft-like" or "airplane-like" operations. The combination of aircraft-like operations and reusability has sparked a revolution in thinking about space transport. There are two ways in which one can think of a space vehicle as operating like an airplane. In one, the most simplistic, obvious, and literal, the vehicle has wings and takes off and lands horizontally like an airplane. It also might use air-breathing jet engines and have a pilot and co-pilot sitting in a cockpit. Because of their conceptual resemblance to airplanes, these space vehicles usually are called aerospace planes and are particularly appealing to the air force.[1]

The second meaning of aircraft-like operations refers to a rocket-powered orbital craft designed to have the same rapid turnaround, ease of maintenance, economy of operation, and abort capability found in the commercial airplane industry. This is the notion of aircraft-like operations that has revolutionized launcher thinking. A rocket-powered, reusable single-stage-to-orbit vehicle with aircraft-like operations is the definition of a true spaceship.[2] True spaceships can launch several dozens (if not hundreds) of times, can fly again shortly after landing (in two weeks or eventually in 24 hours or less), and use small flight and operational teams.

In this sense, the space shuttle is not a true spaceship. Although each orbiter has flown dozens of times, the shuttle system requires a month or more of preparation for the next flight as well as flight and operational teams numbering in the thousands. A true spaceship is a transportation system, like a railroad train or an airplane. When a malfunction occurs during flight, spaceships can abort and land intact, unlike throwaway rockets that have to be blown up. Costs decline because one no longer needs to build and qualify a new rocket for each launch. "Launch" then becomes an antiquated term; "take off" and "land," words used to describe airplane operations, are more appropriate. Making reusable space launchers that have the characteristics of airplane operations is thus the essence of the revolution in thinking of a true spaceship, but it is only part of the space revolution.

Additional parts of the vision involved how to run the program to develop the rocket-powered, single-stage-to-orbit vehicle with airplane-like operations. Such a practical and reality-oriented concern recalls the fact that, although visionaries, the key players were, after all, engineers trained to solve problems in the real world. The vision included the use of experimental flight vehicles known as "X" vehicles and what we shall call here a "faster, cheaper, smaller" managerial approach. The "X" vehicles got their name from the rocket-powered aircraft flown by the air force and the National Advisory Committee for Aeronautics (NACA) and its successor NASA, such as the X-1 and the X-15. They came to stand for a particular way of conducting a program in which one built "X" vehicles in order to flight test technologies incrementally, gradually making increasing demands of the technology and the vehicle. Often referred to as the "build a little, fly a little" approach, it might also be called "build a little, crash a little," because vehicle damage and even loss was an accepted risk when flying an "X" vehicle.

The "faster, cheaper, smaller" (also known as "fast track" or simply "rapid") program management approach stands in contrast to the typical government program. Both "faster" and "smaller" work to keep costs down, which is necessary for a project with a modest budget. Reducing the length of a program keeps costs down, because the longer a program runs, the more it costs just for overhead and labor. Similarly, keeping staff levels at a minimum also decreases costs. Consequently, "faster, cheaper, smaller" projects produce results rapidly at relatively low cost.

The SDIO exemplified this approach; however, its bureaucratic arrangement was actually patterned on a much older bureaucratic practice known as a special projects office. These featured a small staff, a modest budget, and a short timetable for completion. An excellent example of "faster, cheaper, smaller" in action was the air force's ASSET (Aerothermodynamic/elastic Structural Systems Environmental Tests) program. Started in 1959, ASSET cost $21 million plus another $20 million for boosters. From the outset, the size of the project office was minimal. The project manager supervised a staff of only four engineers and one secretary to carry out management, and another seventeen Air Flight Dynamics Laboratory (AFDL) engineers supported the lab's ASSET efforts. However, the industry counterpart, McDonnell, had a larger staff that, at its peak in 1963, consisted of 269 individuals involved with tooling, experimental construction, engineering, and quality control. The McDonnell and AFDL managers worked together closely to minimize bureaucracy and paperwork. Minimal documentation was a goal.[3]

This book does not recount the history of "faster, cheaper, smaller" or "X" vehicles or "aircraft-like operations." Rather it is the story of how they came to be linked to the building of a single-stage-to-orbit rocket in a unified new vision. The novelty of the vision was not its separate elements, but the combination of ideas. This is also the story of how that vision became a government program (the SDIO's SSTO program) and a tangible flying vehicle, the *DC-X*. The program, the vehicle, and the people associated with them convinced many of the feasibility of single-stage-to-orbit transport and, beyond that, of the desirability of the triad "faster, cheaper, smaller," "X" vehicles, and "aircraft-like operations."

Although the technology and hardware of the *DC-X* and single-stage-to-orbit transport are at the center of this story, it is not about the development of single-stage-to-orbit technology or any other kind of technology. Many who knew and scorned the *DC-X* considered it to be a "joke" or "stunt" because the vehicle did not test any new technologies. Developing and testing new technologies was not the point of the program.

In addition to the conceptual elements forming the central single-stage-to-orbit vision, there are elements of politics and political change in this story. It inevitably has a political element because it involves a government program. The essence of politics is the allocation of national resources, especially through the federal budget. This story also deals with the cold war and the impact of the ending of the cold war on a government program. The narrative takes place within the heart of the congeries of institutions and corporations dubbed by President Dwight D. Eisenhower as the military-industrial complex, which arose following World War II to fight the cold war. The ideas of "X" vehicles, applying "aircraft-like operations" to rockets, and "faster, cheaper, smaller" project management all had their origins within the cold war and the military-industrial complex, which the aerospace sector quickly came to dominate. By 1959, two years after *Sputnik I* and the year of NASA's creation, the aerospace industry had become the largest employer in the United States after the automobile industry.[4]

Suddenly, or so it seemed, the cold war came to an end. The Berlin Wall was gone and the Soviet Union ceased to exist, having dissolved into the Commonwealth of Independent States. With the termination of the cold war came deep budgetary cuts, especially to defense spending. Not only would the cost and size of programs have to change, but many programs intended for fighting the cold war would have to cease. The SDIO's SSTO program felt the impact of being associated with the nation's cold war defenses, as well as the reorientation of the Strategic Defense Initiative under President George H. Bush. "Peace dividend" defense cuts saw the SSTO program competing with other launch systems for shrinking budget dollars.

Political change continued to impact the program when the country elected President William Jefferson Clinton in 1992. The new White House wanted its own launcher policy, one distinctly different from that of its predecessors of the opposite political party. As a result, the SSTO program (renamed the Single Stage Rocket Technology Program under Clinton) found itself in a purgatory overseen by congressional funding fights, bureaucratic machinations, unsettled policy issues, and cynics who questioned the feasibility of single-stage-to-orbit transport. Although funded by Congress, the *DC-X* could not fly or pay its bills. As pragmatic as the vision of a single-stage-to-orbit spaceship had been, the tangible expression of that vision had become jammed in the windmills of bureaucracy, national policy, and political change.

In order to understand any aspect of the political history of space over the past two decades, we must consider the space initiatives and policies of President Ronald Reagan. These initiatives set the nation on a path to building a single-stage-to-orbit vehicle, created the agency that would build the *DC-X*, and opened space for commercial exploitation. The eight years of Reagan's reign saw the United States undertake more new space initiatives and more large space initiatives than any of his predecessors since President John F. Kennedy. One day, the 1980s may be remembered as a major turning point in the history of space exploration perhaps as pivotal as the 1960s.

Reagan space policy focused more on space applications than on exploration. Perhaps the most unforgettable Reagan space program was the Strategic Defense Initiative (SDI), a space-based antiballistic missile defense system. Reagan also committed the nation to building a space station by the decade's end, with homage to President Kennedy. Less memorable was the National Aero-Space Plane (NASP), commonly confused with the Orient Express. The Orient Express would have been the nation's fastest aircraft, capable of flying from Washington to Tokyo in two hours, while the NASP would have been the world's first single-stage-to-orbit spaceship. NASP, though, was the antithesis of the true spaceship. The most influential and most lasting of Reagan's space initiatives was the formulation of the first national policy to foster the commercial use of space. The role of the private sector in space has grown tremendously since the end of the cold war and has provided the aerospace industry a respite from the impact of defense cuts.

President Reagan's space programs and policies embodied what is called here the conservative space agenda. The election of Ronald Reagan in 1980 marked the triumph of a national political shift to the Right that started in the 1960s with the presidential campaigns of Barry Goldwater and George Wallace. The conservative space agenda projected the ideas of the so-called New Right into space in the form of policy and programs. Stated simplistically, the conservative space agenda favored business and defense over social issues. The two pillars of the agenda, space commercialization and militarization, rested on a foundation of inexpensive, reliable space transport provided by NASA's space shuttle. This "space truck" would herald a new era characterized by a space-based defense system not just for ballistic missiles, but, for the first time, for the entire civilian population, and the opening of space as a new "frontier" for doing business analogous to the opening of the Western frontier. Since the Reagan era the conservative space agenda has undergone transformation, but the country has continued its political drift to the Right.

This, then, is the story of a pragmatic vision of reusable, rocket-powered, single-stage-to-orbit spaceships. Our story includes an examination of how this vision materialized as a government program and an experimental vehicle and how that program was run. We will look at the separate elements of the vision and how they came together into a single vision, as well as the process of using political leverage to turn that vision into a government program. Along the way, we shall see how both domestic and international politics and political changes aided and hindered the program.

The history consists of four parts. The narrative is largely nonlinear and often alternatively stochastic and parallel in a non-Euclidean sense. Reality is more complex and more random than the linearity and predictability of traditional narrative. A chronology is therefore included to provide readers a series of guideposts in the form of program milestones and historical events.

Part I (chapters 1, 2, and 3) examines the conservative space agenda as expressed and carried out by the Reagan administration. Chapter 1 considers the political and intellectual underpinnings of the conservative space agenda, from the country's shift to the Right to the Reagan Revolution, and from Newt Gingrich to the space version of Alfred Mahan. It concludes with a discussion of the Reagan administration's commercial space policy, which assumed a symbiotic relationship between commercial and military interests, and which drew inspiration from the low-cost access to orbit promised by the shuttle. Chapter 2 investigates the various efforts of the Reagan administration to jumpstart space commerce, with special attention to the roles of the shuttle and Arianespace competition. Chapter 3 tracks the development of the Strategic Defense Initiative (SDI) and its promotion by such cold warriors as Lt. Gen. Daniel O. Graham, with an emphasis on how SDI differed from earlier missile defense and antisatellite systems.

Part II (chapters 4 and 5) reviews the ongoing quest of space visionaries to create a single-stage-to-orbit vehicle. Chapter 4 discusses the notion of "aircraft-like operations" and traces the idea of the aerospace plane from its beginnings to the National Aero-Space Plane of the 1980s. Although intended to be a single-stage-to-orbit transport with "aircraft-like" operations, NASP was a typical cold war, large-scale, long-term, expensive, heavily bureaucratic project. Chapter 5 starts with early concepts for rocket-powered, single-stage-to-orbit transport and traces the development of a new approach by space visionary and cold warrior Maxwell W. Hunter II. To the reusable single-stage-to-orbit vehicle, he added "aircraft-like operations" and the belief that it should be developed incrementally, using an "X" vehicle in a "faster, cheaper, smaller" program.

Part III (chapters 6 and 7) recounts how this fresh vision of single-stage-to-orbit transport, dubbed the *Space Ship eXperimental (SSX)*, became an "X" vehicle, the *Delta Clipper Experimental (DC-X)*, in a government program. Chapter 6 underscores the role of the space movement and proponents of the conservative space agenda in furnishing the single-stage-to-orbit prophets access to the White House, the first step in turning the vision into a government project run by the Strategic Defense Initiative Organization (SDIO). Chapter 7 recounts how the SDIO managed the program from the Phase I design studies to construction of the Phase II vehicle, as well as the process by which McDonnell Douglas built the *DC-X*. The chapter concludes with observations on a number of the changes that took place as the *DC-X* actualized the *SSX* vision: the end of the cold war, the transformation of the Strategic Defense Initiative, and the introduction of "smaller, faster, cheaper" management at NASA. These changes hinted that NASA, rather than the SDIO, might be a more appropriate institutional home for the *DC-X*.

Part IV (chapters 8, 9, and 10) relates the struggle to keep the program funded and alive in the midst of budget cuts, funding fights with other launch programs, bureaucratic maneuvers, charges of technological unfeasibility, and yet another political sea change. In chapter 8, as the *Delta Clipper* sailed through the tempestuous waters of change that marked the years immediately following the cold war's end, the role of the network of friends and supporters of the *DC-X* and single-stage-to-orbit transport became critical. The chapter recounts their struggles against budget cuts, competing launch systems, and claims supported by a NASA study that single-stage-to-orbit was technologically unattainable until sometime in the distant future. It ends with the search for a new agency to fund and manage the project.

Chapter 9 deals with events that occurred following the election of President Clinton. His calls to make government more "businesslike" and to cut waste, red tape, and spending harmonized with the updated conservative space agenda. The chapter discusses the unsuccessful efforts to save Phase III, the first flights of the *DC-X*, and fresh attacks in Congress. It also shows how a small, but influential, group of NASA officials became convinced of the feasibility and desirability of single-stage-to-orbit rockets.

The last chapter reviews NASA's emerging interest in single-stage-to-orbit transport and shows the impact of Clinton administration launch policy on the *DC-X* program, namely, its transfer to NASA. As a result, the first peacetime presidency following the cold war decided that the *DC-X*'s institutional home would not be within the military-industrial complex, but with the civilian space agency. The chapter ends with an account of the vehicle's final flights as NASA's *Clipper Graham*, named, ironically, for cold warrior Daniel Graham.

Part I / The Conservative Agenda for Space

CHAPTER 1

The Reagan Revolution

Our story begins in the 1980s, during the last years of the decades-long cold war, a war that profoundly transfigured the cultures of North America and Western European countries and the lives of their inhabitants. The presidency of Ronald Reagan marked a sea change in the country's conduct of the cold war and its space activities, as well as the triumph of America's political turn to the Right. Space policy held a high place during the Reagan administration. Ronald Reagan's two terms as president saw the United States undertake more new and more large space initiatives than any previous administration since that of John Kennedy.

Conscious of the comparison, Reagan often styled himself after Kennedy in his space policy speeches. For example, during his 1984 State of the Union address, Reagan attempted to emulate the drama and boldness of Kennedy's challenge to land an astronaut on the Moon by the end of the decade. He "dared" America "to be great." "We can follow our dreams to distant stars, living and working in space for peaceful, economic, and scientific gain. Tonight," he announced, "I am directing NASA to develop a permanently manned space station and to do it within a decade."[1] Like Kennedy, Reagan would pass on a lasting legacy to the nation's space program.

In addition to the space station, Reagan space projects included the Strategic Defense Initiative (SDI), a space-based antiballistic missile defense system, and the National Aero-Space Plane. The most influential and most lasting of Reagan's space projects was the formulation of the first national policy on the commercial use of space. These programs and policies very likely will cause the 1980s to be remembered as a major turning point in space history as important as, or possibly even more important than, the 1960s.

The course of the cold war and America's military uses of space changed dramatically following Reagan's election. Foreign policy went from détente to hostility. SDI was a radical shift in military thinking about space that, unlike previous space defenses, placed weapons in orbit. Equally radical was Reagan's policy to open space for commercial ventures. American telecommunication firms had been exploiting space for business purposes, mainly long-distance communications, since at least the 1950s. However, the Reagan administration was the first to pursue the stimulation of space business with vigor and to make it the subject of national policy.

Reagan space initiatives and policies mirrored the political ideas that had helped to bring him to power. His election in 1980 (and, to a lesser extent, the takeover of the Senate by the Republican Party) brought to power in Washington a new group of policymakers who sought to remake virtually every federal policy and program. Some of the new arrivals had definite ideas about how the nation should use space. Those ideas constituted what is called here the conservative space agenda.

The essence of this agenda was to project into space political ideas then current among many members of the New Right as well as traditional Republican beliefs. Expressed simplistically, the conservative space agenda favored business and defense over social issues. It shared the New Right's antipathy toward the triumphs of the Left, notably President Lyndon Johnson's Great Society programs, the civil rights movement of the 1960s, and the ongoing "rights revolution," which included not just blacks and Hispanics, but women and gays as well. The conservative space agenda expressed a fervent patriotism, strong anticommunism, and disgust for the "welfare state" and "big government." The New Right's alarm over the state of the U.S. economy and industry's lack of international competitiveness also found itself in the conservative space agenda, as did calls for the deregulation of commerce, cutting the size of government, and reductions in corporate and individual tax rates. A belief in the positive benefits of technological progress and scorn for the notion of limits to growth also marked conservative space agenda supporters.

What inspired and empowered the conservative space agenda was a new form of space transport, NASA's space shuttle. Before it began to fly in 1981, the shuttle appeared to herald a new era of inexpensive, routine space travel. Thanks to the "space truck," business ventures would play a much larger role in space, and space-based defenses would protect both military assets and the civilian population. The shuttle also embodied the period. The exploratory nature of the nation's space program gave way to the exploitation of space. As a launcher of Earth-orbiting civilian, military, and commercial payloads, and as a platform for conducting experiments in near space, the shuttle was the emblem and embodiment of this new exploitive space program that concentrated more on commercial and military space applications than on space exploration.

The conservative agenda was about more than the shuttle, or SDI, or the commercial exploitation of space. It had a conceptual side that was as genuine as any offered by the New Right. In order to understand the conservative space agenda more completely, and how it mirrored the New Right, we will look first at some of the political ideas and economic conditions that helped bring about the so-called Reagan Revolution. Next, we will examine the writings of the leading spokesperson for the conservative space agenda, Newt Gingrich, and follow up with some ideas of two of our story's chief protagonists, Lt. Gen. Daniel O. Graham and Maxwell W. Hunter, II.

The Reagan Revolution

Reagan's election in 1980 marked the culmination of a larger political shift to the Right that had its roots in opposition to the Great Society programs and the civil rights movement of the 1960s. Barry Goldwater's unsuccessful 1964 presidential campaign was the first to express this conflict, and the presidential candidacies of George Wallace and Richard Nixon later capitalized on it. By the 1968 elections, antipathy among white Americans had hardened into a growing fear of black street crime, disgust with urban riots, resentment of student rebels, and opposition to perceived special treatment for blacks.[2]

America's shift to the Right also drew on a general dissatisfaction with the state of the U.S. economy. "Stagflation," the combination of simultaneous stagnation and inflation, continued to trouble the country throughout the 1970s. Productivity declined. The trade deficit skyrocketed from $9.5 billion in 1976 to $31.1 billion in 1977 and to over $34 billion in 1978. At the same time, the economies of Singapore, Japan, South Korea, and Taiwan were expanding rapidly and seeking new export markets, which they found in the United States. Traditional American trading partners in Europe also were eager to expand into the U.S. market. As a result, key American industries, such as automobiles, steel, and electronics, quickly faced a deluge of quality goods from abroad. In order to compete, industry started moving plants and jobs to either the Sunbelt or the Third World in search of cheaper labor and operating costs.[3]

Although the business cycle initially appeared to work in favor of President Jimmy Carter, in June 1979, OPEC nearly doubled its price of crude oil. The chairman of the Federal Reserve Board, Paul Volcker, instituted a series of politically insensitive, draconian measures to cut inflation and protect the dollar abroad. The elevated interest rates and reduced money supply induced a recession in 1980, a presidential election year.[4]

As a remedy for the unhappy state of the economy, the New Right advocated reducing corporate and individual tax rates and cutting the size of government. Americans were in search of tax relief and already, in June 1978, California voters had approved Proposition 13, which cut property taxes.[5] The New Right's antigovernment and probusiness positions received theoretical support from supply-side economic theory and the famous curve of economies professor Arthur Laffer. If tax rates were too high or too low, Laffer argued, they would neither produce economic growth nor generate sufficient government revenues. The essential prescription of these theoretical arguments was to cut taxes for business and individuals and to reduce the size of government bureaucracy.

Within the first two years of his presidency, Reagan, along with a Congress with a Democratic majority in the House, cut taxes, rolled back many Great Society programs, and weakened the regulatory authority of the federal government. The Economic Recovery Act of 1981 (ERTA) fulfilled the supply-siders' dream by dropping the top individual tax rate from 70 to 50 percent, cutting corporate taxes, and giving tax breaks to key industries. The individual tax rate cuts helped mostly those with the highest incomes, freeing up money for investment. It alleviated the tax burden on those with median incomes only slightly (3.5 percent), while actually increasing the tax bite on families with the lowest annual incomes. Corporate America received a hefty $150 billion tax cut over a five-year period. By 1986, ERTA had deprived the federal government of an estimated $600 billion in tax revenues. In addition, the Omnibus Budget and Reconciliation Act (OBRA) of 1981 cut over $35 billion from the federal budget for fiscal 1982 and a total of $140 billion through fiscal 1985. OBRA went far toward achieving the New Right goal of rolling back Great Society programs by cutting or abolishing support for the most politically unpopular social programs of the 1960s War on Poverty.

Building on the foundation laid by Jimmy Carter, who had deregulated the airline and trucking industries, Reagan weakened a number of key regulatory agencies that oversaw business practices. In his words, government had "overspent, overestimated, and overregulated." Among the affected agencies were the Occupational Safety and Health Administration, the Environmental Protection Agency, the Federal Trade Commission, and the Federal Communications Commission. Reagan also approved legislation freeing the savings and loan industry to move into commercial banking. Later, the country would have to pay a high price for that piece of deregulation.

These regulatory, funding, and tax innovations did not immediately provide the business community a tangible basis for its euphoria over the election of Ronald Reagan. The country was still in a recession. Nonetheless, as the inflation rate fell, as the OPEC cartel's oil prices collapsed, and as deregulation and ERTA and OBRA legislation went into effect, the economy began to turn around starting in 1983. The country now witnessed a dynamic upsurge of entrepreneurship and a new mood of optimism in the business community. As a result of tax relief for corporations and wealthy individuals, more money was available for investment in new and existing enterprises, an estimated $120 to $160 billion per year.[6] This flood of venture capital promised to buoy the emerging U.S. commercial launch industry and to bankroll business ventures in space. Typifying the optimism of the era was a statement by Norman R. Augustine, then president of Martin Marietta Denver Aerospace and a former undersecretary of the army. He told attendees at a conference on the future of space: "First of all, let me say that we're terribly fortunate that we have the strong support of our President for space commercialization."[7]

Space: The Newt Frontier

Deregulation and tax cuts as business policy had their equivalents in both Reagan space policy and the conservative space agenda, as did other aspects of New Right (and traditional Republican) ideology, such as an abhorrence of welfare and a desire to prosecute the cold war aggressively. One of the most influential and vocal proponents of the conservative space agenda was Newt Gingrich. Permeating his thinking about space was an implicit enthusiasm for technological progress and a disdain for the idea of limits to growth.[8] The oil embargoes of the seventies and the resulting high cost of petroleum-derived energy had encouraged some to reconsider how industrial societies use natural resources. The search for alternative renewable fuel sources, the institution of programs encouraging recycling and reuse, and the enactment of other conservation measures made sense in the face of what appeared to be limits to industrial growth and finite natural resources. For the most part, however, the average American believed that the major oil companies had created the oil shortage in order to reap enormous profits. In the fall of 1979, oil shortages began to ease, and by October, the odd/even system of gas rationing had ended everywhere, fueling new optimism. That optimism spread to the thinking about space. Space became a new frontier, like the fabled frontier of the Old West, where new resources and new (business) opportunities abounded. There were no limits in space save those of the imagination.

Newt Gingrich was not only an avid supporter of the *DC-X* program, but he also was a member of Congress and had access to President Reagan. Few others have written as much, or as coherently, as Gingrich on what (part of) the New Right wanted to do in space. His space agenda brought together ideas from the New Right as well as from futurology and science fiction. Sharing and helping to shape his ideas on space were individuals who also had the ability to influence Reagan space policy. In fact, Gingrich's ideas about space are less important as a reflection of his personal thinking than as a mirror held up to reflect the thoughts of a number of like-minded individuals.

Elected to the House of Representatives from Georgia for the first time in 1978, Gingrich began formulating his ideas about the future, space, and technology in late 1982 and early 1983. The economy was beginning to turn around, and a mood of optimism was spreading among conservatives and the public in general.[9] In a book (*Window of Opportunity*) and an article, both published in 1984, Gingrich described a future world marked by the achievement of New Right goals, a world he called the Conservative Opportunity Society. The exploitation of space was an essential part of that world.

At the heart of Gingrich's futuristic vision was a three-pronged attack on the notion of limits to growth, a future of lowered expectations, and the need for state control and planning. Gingrich attributed these ideas to liberalism. "We must first destroy," he wrote, "this liberal cultural and psychological myth of limits."[10] He sought to imagine a space program that would change the economy, society, politics, and culture, just as the discovery of the New World had changed Europe. "Breakthroughs in computers, biology, and space," he declared, "make possible new jobs, new opportunities, and new hope on a scale unimagined since Christopher Columbus discovered a new world."[11] For Gingrich, technology and space were a fundamental part of the American ethos, the frontier spirit.

In *Window of Opportunity,* Gingrich combined technological optimism with attacks on welfare and "big government," but with a twist. "This optimistic future," Gingrich explained, "will require reforms—both in theory and practice—of the welfare system." The handicapped will no longer be dependent on welfare, but will find gainful (tax revenue–generating) employment thanks to new technologies ("compassionate high tech") and scientific discoveries.[12] "Compassionate high tech" essentially argued that the benefit of investing in commercial and military space technology (in fulfillment of the conservative space agenda) would "trickle down" to Earth and lighten, if not resolve, the need for social welfare in a technology-oriented version of "trickle down" economics.

One of the most significant scientific and technological fields in which Gingrich's future would take root was space. The space station, endorsed in President Reagan's 1984 State of the Union message, was an important part of that future. Gingrich's space station would have "a biology and health care module, a chemistry and manufacturing laboratory, and a service bay in which satellites could be repaired by crews wearing spacesuits, as well as a resources module to provide the energy and life-support systems for the four human-habitat modules." The station would serve scientists and environmentalists, as well as an array of industrialists, from those interested in medicine, crystals, semiconductors, new alloys, and optical fibers to those in the satellite communications business. He also called for creation of a permanent lunar base, housing at first "only four to six people, probably on a three- to six-month rotating cycle much like the regime in Antarctica." The lunar base would be "the most appropriate single millennium project . . . opening in January 1, 2000 . . . for the whole free world."[13]

Although that base might be in the Sea of Tranquility, Gingrich did not see tranquility for his Conservative Opportunity Society in space. The Soviet Union threatened it. Cold war rivalry in space would continue into the future. The United States had to hasten its plans for a space station, he warned, because the Soviet Union was working on a space station, too.[14]

America needed to increase NASA's budget, Gingrich argued, because the country simply did not spend enough money on space for any benefits to materialize. NASA's annual budget would run the Defense Department for only eleven days or Health and Human Services for only eight days. Over thirty corporations—including such NASA contractors as RCA, General Electric, IBM, Westinghouse, and Western Electric—were larger than NASA. "The misconception of the size of NASA's budget," Gingrich explained, "is due in part to the effect of inflation." When viewed in constant dollars, NASA's budget was less than half of its peak 1965 budget, he pointed out. The agency's budget had to be increased significantly. A large NASA budget was not inconsistent with conservatives' stand against "big government." "Conservatives are not against a strong government," he explained. "Conservatives are against big, bureaucratic welfare states."[15]

Gingrich proposed "five simple steps to a bold future." The first was to spend as much on civilian as on defense space ventures. The second was to "make a national commitment" to keep NASA's budget at its peak 1965 level (adjusted for inflation) through the year 2000. Echoing New Right economics, he also recommended changing the tax and regulatory environment in order to commercialize space. "It is in our long-term interest as a nation to create a climate in America conducive to the exploration of space—the only frontier—which encourages the risk-taking and daring that only private initiative can risk. The longer we keep space a government monopoly, the greater the danger of smothering initiative." His fourth proposal was to involve the country's allies and friendly developing nations (especially in Latin America) in exploiting space. This would "increase our prestige more than trillions in foreign aid." Finally, Gingrich recommended cultivating "the beginnings of populism in space and the beginnings of citizen involvement by initiating a tourism program for the American people." This involved sending ordinary able-bodied citizens, selected by "lottery based on individual income tax forms," into space on NASA's shuttle. "More than anything else," Gingrich reasoned, "this would tie the average American into the space program."[16]

If Gingrich's writings sound like those of a futurist or a science fiction aficionado, it is no coincidence. First among Gingrich's book acknowledgments in *Window of Opportunity* is Jerry E. Pournelle, science-fiction writer and senior consulting editor at *BYTE* magazine.[17] Pournelle, known in certain circles as a visionary and a good communicator about space,[18] also had a solid knowledge of defense and aerospace technology. In 1970, for example, he and Stefan T. Possony, a Pentagon intelligence officer and Georgetown University professor of international relations (later a senior fellow at the conservative Hoover Institution), wrote *The Strategy of Technology*. In one chapter, "Assured Survival," they outlined a strategic substitute for Mutual Assured Destruction.[19] Pournelle earlier had been associate director of operations research at North American Rockwell's Space Division, as well as president of the Pepperdine Research Institute, which conducted studies on technology development and strategy for the air force. As an employee of the Aerospace Corporation, Pournelle performed general editorial and systems analysis work for the U.S. Air Force's Project 75, a classified study to project ballistic systems technology needs to the year 1975.[20] Pournelle became well connected to conservative political circles in California, while writing speeches for then-Governor Ronald Reagan, and in Washington, after Reagan's election to the presidency.

Pournelle was not the only science fiction writer on whom Gingrich drew for ideas in *Window of Opportunity.* He consulted the renowned Ben Bova and the lesser light G. Harry Stine, who played a minor part in our story.[21] Gingrich clearly owed an additional debt to certain futurist writers, such as Herman Kahn[22] and Alvin Toffler, author of *Future Shock* and *The Third Wave,"*[23] especially to Toffler's idea of an emerging information society. Indeed, Gingrich's *Window of Opportunity* had such a futurological approach that he published an abbreviated version of it in the journal *Futurist.*[24]

Gingrich's co-author of the article, "Space: The New Frontier," was James A. M. Muncy, a space activist with a low-level White House connection. In 1979, while still a college student, Muncy had organized a grass-roots group, the Action Committee on Technology, to lobby for legislation favorable to creation of a high-tech society. Then, after working as an intern under Gingrich for two and a half years, he served with President Reagan's science advisor, George A. Keyworth II, working primarily on space issues.[25] Thus, although Gingrich's ideas drew upon science fiction writers and futurists, they also had a direct link to policy formulation within the White House, as well as expression on the floor of the House of Representatives.

Mahan in Space

In addition to the ideas expressed by Newt Gingrich, certain ideas held by Max Hunter and Daniel Graham are relevant, too. Both men argued in favor of a space-based defense system. For the most part, their thinking illustrates the interdependency of commerce and defense that was at the heart of the conservative space agenda. Promoting defense benefited commerce, and the promotion of commerce benefited defense. Graham and Hunter believed that a U.S. space–based global defense system like SDI would bring about a Pax Americana similar to the Pax Britannica induced by Britain's domination of the world's oceans.[26]

The metaphor linking the oceans and space was neither new nor unique. For example, in a speech delivered September 24, 1962, President Kennedy told a crowd at Rice University stadium: "We intend to be first . . . to become the world's leading space-faring nation."[27] While the term "space-faring" may have made the president sound like a science fiction writer to some, it rounded out the speech's sustained allusions to the seas and exploration. For Graham and Hunter, the benefits of dominating space, like the benefits of dominating the seas, would be both military and commercial.

Graham drew on the theories of Alfred T. Mahan (1840-1914), a famous naval strategist and professor at the Naval War College. Mahan stressed the interconnection between the commercial exploitation of the oceans and the military advantages of dominating the seas.[28] He based his beliefs on studies of the role played by control of the sea, or the absence thereof, in the course of history up to the Napoleonic Wars.[29] "Let us start from the fundamental truth, warranted by history," Mahan wrote, "that the control of the seas, and especially along the great lines drawn by national interest or national commerce, is chief among the merely material elements in the power and prosperity of nations."[30] Later he wrote that the sea is: "the most potent factor in national prosperity and in the course of history."[31] As with New Right and traditional Republican thinking, Mahan's theories made encouraging commerce a fundamental priority. Moreover, like them, Mahan saw no difference between "national interest" and "national commerce."

Commerce was "the national characteristic most important in the development of sea power." Sea-borne commerce made a nation great. Therefore, it was necessary for nations to provide security to maritime commerce. Military command of the sea would assure the strength needed to win a war. Command of the sea during war or peace also guaranteed the continuance of maritime commerce. For Mahan, sea power was not synonymous with naval power. It included the military strength that ruled the sea or any part of it by force of arms as well as "the peaceful commerce and shipping from which alone a military fleet naturally and healthfully springs, and on which it securely rests."[32] In this way, Mahan affirmed the close relationship between foreign trade and the navy.

Mahan sought to extend the commercial and military influence of the United States in the Gulf of Mexico and the Caribbean Sea, especially at the Isthmus of Panama, where a French company planned to build a canal.[33] Similarly, the conservative space agenda anticipated expanding U.S. military and commercial activities into Earth orbit. Mahan believed that the commercial and military importance of the oceans derived from the fact that: "the sea is the world's greatest medium of circulation."[34] Space had not yet (and still has not) achieved the same level of activity or commercial value as the oceans.

Reagan Revolution in Space

What is important to recall about Graham's use of Mahan is the relationship it established between commerce and defense. The conservative space agenda would extend this relationship into space. The Reagan administration set out to accomplish (consciously or not) the conservative space agenda and heralded a new space era. It took defense to the "high ground" (the "high frontier") by inaugurating the first space-based defense system, the Strategic Defense Initiative, and it took measures to foster the development of space business, which was still in its infancy. At the highest levels of government, the formulation of commercial space policy went hand in hand with military space policy. Not surprisingly, Reagan commercial space policy bore a few marks of military influence; however, for the most part, it mirrored the values of the New Right and the conservative space agenda. Reagan policy particularly sought to create a climate conducive to investing and doing business in space. Typical of the space commercializing ventures of the eighties, the shuttle strongly influenced Reagan policy.

Political scientist W. D. Kay has pointed out that "for the first several months of his presidency, Ronald Reagan did not appear even to have a science policy of any sort, let alone a plan for the U.S. space program." That changed after the first flight of the shuttle in April 1981. The general feeling in the White House was that anything was possible. The shuttle, Kay argues, "appeared to provide the Reagan White House with the final ingredient—the requisite technology—that it needed to integrate the U.S. space program into its larger political and economic goals."[35]

Reagan commercial space policy grew out of an examination of military space policy carried out at the highest level. In August 1981, the president directed the National Security Council to review space policy. A senior interagency group, known as SIG (Space), came together under the direction of the president's science advisor, George Keyworth, and the National Security Council. The group's chair was Victor H. Reis, an assistant director of the Office of Science and Technology Policy. Agencies participating included the Departments of State, Defense, and Commerce, the CIA, the Joint Chiefs of Staff, the Arms Control and Disarmament Agency, NASA, the National Security Council, and the Office of Management and Budget. The interagency group addressed launch vehicle needs, the adequacy of existing space policy for national security needs, shuttle responsibilities and capabilities, and potential legislation.[36]

The National Space Policy, issued as National Security Decision Directive 42 (NSDD 42) on July 4, 1982, included business in space policy for the first time and marked the start of a national policy on space commerce. The idea of economic benefits from space (such as telecommunications, weather forecasting, remote sensing, and navigation) was not new; however, this was the first time in the history of the U.S. space program that a high-level official document made a direct reference to the American business community. Nearly all preceding space activity had been undertaken by governments. NSDD 42 thus marked a dramatic redefinition of space policy not seen since the launch of *Sputnik I* in 1957.[37] NSDD 42 set four goals to be accomplished in space: the first two dealt with national security, while the third and fourth called for "obtain[ing] economic and scientific benefits through the exploitation of space" and for "expand[ing] United States private-sector investment and involvement in civil space and space-related activities."[38]

The release of the 1982 National Space Policy revealed its indebtedness to the shuttle. NSDD 42 called for making the shuttle available to all commercial users, provided no conflicts with national security resulted. On July 4, 1982, the same date as the release of NSDD 42, President Reagan spoke before an audience of some fifty thousand people at Edwards Air Force Base, with American flags flying in the background, as the *Columbia* landed.[39] This was the shuttle's final test mission and the beginning of its operational status. It also was the first mission to carry a Pentagon payload and the first "Get-Away Special" experiments conducted for a NASA customer.[40] Southern California was a politically astute locale for proclaiming support for commercializing space. Much of the nation's aerospace industry was located there. As the state's former governor knew, the California aerospace industry was, in the language of an internal note, "very depressed" with "80,000 unemployed."[41] Commercializing space very much concerned people on Earth.

The Space Launch Policy Working Group addressed the role of the shuttle in the emerging commercial launch market in an April 13, 1983, report. The same report also considered competition from overseas, U.S. commercial launchers, and government launch needs. The Space Launch Policy Working Group was an eleven-member committee that consisted of representatives from the Office of Management and Budget, the Arms Control and Disarmament Agency, the Joint Chiefs of Staff, the Office of Science and Technology

Policy, the Central Intelligence Agency, the Department of State, the Department of Commerce, the Department of Defense, and NASA. Charles Gunn, NASA, and Thomas Maultsby, Defense, jointly chaired the group.

The Space Launch Policy Working Group recommended that the shuttle be made available to all "authorized users," domestic or foreign, whether commercial or governmental. The government had to consider the impact of NASA's pricing of commercial and foreign shuttle payloads on the commercial launch industry. Nonetheless, the working group advised, "the price for commercial and foreign flights on the STS [shuttle] must be determined based on the best strategy to satisfy the economic, foreign policy, and national security interests of the United States."[42] Thus, commercial and defense considerations would be taken into account in conjunction with one another.

On May 16, 1983, the president issued a new National Security Decision Directive that reflected the Space Launch Policy Working Group report and Reagan's own thinking.[43] Its central objective was to encourage the U.S. commercial launch industry. The policy would make the shuttle available to all domestic and foreign users, whether governmental or commercial, for "routine, cost-effective access to space." It also would promote the commercial use of expendable rockets by making government ranges available for commercial launches at prices "consistent with the goal of encouraging" commercial launches, and by encouraging competition "within the U.S. private sector by providing equitable treatment for all commercial launch operators."[44] Already, a fatal flaw was developing within Reagan commercial space policy. Placing commercial payloads on the shuttle while simultaneously encouraging expendable launchers proved to be a paradoxical policy that would stymie the creation of a domestic launch industry for nearly a decade.

On July 20, 1984, the White House released a special National Policy on the Commercial Use of Space. A year before formulating this policy (July 28 and August 3, 1983), White House senior officials had met with representatives from a range of companies interested in conducting business in space. The officials included Edwin Meese III, chief counsel; George Keyworth, the president's science advisor; Craig L. Fuller, assistant to the president for cabinet affairs; James M. Beggs, NASA administrator; and Clarence J. Brown, deputy secretary, Department of Commerce. Among the companies represented were E. F. Hutton, Federal Express, and John Deere & Company, as well as start-up space companies, such as Orbital Sciences Corporation and Space Industries, Inc., and established aerospace firms, including Fairchild Space Company,

McDonnell Douglas Astronautics, Grumman Aerospace, General Dynamics, and Rockwell International.

Industry expressed a variety of views that showed both consensus and conflict among the companies intending to develop space as a place to do business. They agreed mostly on what government could do for business. For example, several firms agreed on the desirability of incentives, such as tax credits (including research and development tax credits), liberal depreciation allowances, low-cost capital, risk sharing, government contributions to an insurance pool, liability caps set by the government, and government loans. Other firms wanted NASA to play a larger role either by conducting research, building a space station, or making the shuttle, as Fairchild wanted, "a common carrier to and from space as rapidly as possible." Only two firms wanted government, namely NASA, to play a smaller role. General Dynamics objected to the government's proposal to maintain current shuttle pricing policy through fiscal 1988, because it would hurt the private sector's ability to compete with the Europeans. Space Services, Inc., wanted the White House to "eliminate governmental competition with private sector space applications wherever possible (particularly when the government competition is subsidized by taxpayer dollars)."[45] Both Space Services, Inc., and General Dynamics already perceived the flaw in Reagan commercial space policy.

The National Policy on the Commercial Use of Space, released on July 20, 1984, provided the same incentives recommended by those firms that had visited the White House. It set out a series of initiatives that included research and development tax credits, a ten percent investment tax credit, accelerated cost recovery, timely assignment of radio frequencies, and protection of proprietary information. In addition, the policy established the Working Group on Commercial Use of Space of the Cabinet Council on Commerce and Trade (CCCT), thus placing space commercialization at the highest level of government. It would include all of the agencies and departments represented in the SIG (Space), which continued to formulate national space policy, as well as all other interested agencies. NASA and Commerce Department representatives would chair the CCCT working group.[46]

Governing the New Frontier

One of the policy questions to be settled was the regulation of the new industry. Ironically, an administration that sought to *deregulate* industry would decide the issue. Also ironic was the fact that the issue had been raised first not by the White House, nor by any government agency, but by a private company, Space Services, Inc. (SSI). The firm wanted to launch its own rocket, the *Conestoga I,* from a private launch site off the Texas coast. SSI tried to clear the launch with the Federal Aviation Administration, the navy, the state of Texas, NASA, NORAD, and the State Department, among others. Each agency (excluding NASA) supported its jurisdiction through existing legislation, since regulations or laws governing private launches did not yet exist.[47]

The Department of State, in response to SSI's request, chaired an ad hoc interagency group to determine the proper requirements and procedures. They concluded that "existing laws and regulations are adequate for the USG [U.S. government] to control private launches from US territory or export of satellites and launch vehicles generally from the US for launch abroad. In light of the infancy of the private space launch industry, the creation of new legislative and regulatory framework and supporting bureaucracy is not justified." The ad hoc group determined that commercial launch firms would have to obtain three basic approvals for a launch: 1) an Arms Export Control license from the State Department, as dictated by the Arms Export Control Act and the International Traffic in Arms Regulations; 2) an experimental radio license from the Federal Communications Commission, in accordance with the Communications Act of 1934 and amendments, and; 3) an exemption or clearance from the Federal Aviation Administration for use of controlled airspace, as mandated by the Federal Aviation Act of 1958.[48]

Within the White House, the Space Launch Policy Working Group created by SIG (Space) recommended that the Department of State serve as an interim point of contact between government and industry. The Department of State also would chair an interim interagency group composed of members from SIG (Space) and observers from other interested agencies, such as the Federal Communications Commission. The proposed group would "streamline the procedures" used "to implement existing licensing authorities and to develop and coordinate the requirements and process for the licensing, supervision, and/or regulations applicable to routine commercial launch operations from commercial ranges."[49]

Reagan had an agency other than the State Department in mind, though. On November 18, 1983, after a meeting of the Cabinet Council on Commerce and Trade, he designated the Department of Transportation (DoT) as the lead agency to "promote and encourage commercial ELV [expendable launch vehicle] operations in the same manner that other private United States commercial enterprises are promoted by United States agencies." Rather than emulate the regulatory agencies scorned by the New Right, hampering commerce and inflating consumer prices, the DoT would "make recommendations . . . concerning administrative measures to streamline Federal government procedures for licensing of commercial" launches. The DoT also would issue those licenses. The agency would "identify Federal statutes, treaties, regulations and policies which may have an adverse impact on ELV [expendable launch vehicle] commercialization efforts and recommend appropriate changes to affected agencies and, as appropriate, to the President."[50] Reagan announced the new role of the DoT in his State of the Union Address on January 25, 1984.[51] Here was a regulatory mandate to encourage industry. Space commercialization was becoming a model of the Reagan Revolution and the conservative space agenda.

Subsequently, Congress gave the Department of Transportation's new role a legal basis with the passage of the Space Launch Commercialization Act, H.R. 3942 (Senate bill S.560). Members of Congress felt that the designation of a lead agency was insufficient because it lacked "legislative authority. The result could inhibit decisionmaking and interagency coordination and allow the present inefficient approaches to commercial launch approvals to persist."[52] The bill passed and became known commonly as the Commercial Space Launch Act of 1984.[53] Acting on the authority of both the act and a presidential executive order, Secretary of Transportation Elizabeth Hanford Dole established the Office of Commercial Space Transportation (OCST). The act stipulated that no U.S. citizen could operate a launch vehicle anywhere, under any circumstances, without a license issued by the DoT. It also excluded NASA and the Department of Defense from licensing requirements. This exclusion included the shuttle even when it carried commercial payloads, thereby giving the shuttle another advantage over potential commercial launch providers.

CHAPTER 2

Commerce on the High Frontier

The space shuttle was one of the foundation stones of Reagan-era space policy. It empowered and made possible that policy by carrying military, civilian, and commercial payloads into orbit. It also inspired entrepreneurs to imagine new space commercial ventures. The Reagan administration offered more than the shuttle to advance space commercialization. Its space policy provided a framework and a regulatory agency, as well as incentives, but not subsidies, for space businesses. This approach was consistent with traditional conservative economics that favored laissez-faire measures. Other measures for fostering space business ranged from privatization to NASA commercialization projects.

The Reagan administration's efforts produced mixed, largely disappointing, results, such as the failed space station. The goal of Reagan space policy, however, was not to get government more involved in space commerce, but to stimulate entrepreneurship. The shuttle's promise of inexpensive, reliable transport stimulated the imaginations of space entrepreneurs and investors, while some firms hoped to create rival, less expensive expendable launchers for commercial payloads. The results, again, were disappointing. Any firm hoping to sell launch services had to contend with the government's carrier, the space shuttle, as well as with the European competitor, Arianespace. Thus, while Reagan space policy aspired to foster the commercialization of space, it placed commercial launch companies between the Scylla of the shuttle and the Charybdis of Arianespace. As a result, too, any firm hoping to exploit a space-based payload had to look to either the shuttle or Arianespace for a ride into orbit. The Reagan agenda to stimulate a commercial launch industry was a failure, and it took a failure of a different, more tragic sort to fix the problem.

NASA Goes Commercial

A key Reagan tool for promoting private ventures in space was government itself through privatization (Landsat) and NASA. At times in U.S. history, the government has intervened actively to support and encourage the economic progress of various endeavors through privatization. There are two basic approaches to privatization. In "load shedding," the government sells, or transfers in some other fashion, a publicly owned asset to a private interest. An example would be the 1987 sale by the Department of Transportation of Conrail, a freight railroad that the government had taken over from the bankrupt Penn Central ten years earlier. The other major privatization arrangement is contracting, in which a private company provides a government agency a good or service under a fixed-term contract. A common example is the moving of federal agencies, such as NASA Headquarters, from a government-owned building into spaces leased from a private office complex.[1] Of the various Reagan administration privatizations, only one, Landsat, dealt with space.

Landsat was a remote-sensing operation run strictly by and for the government. NASA created the Landsat satellites in order to collect different kinds of data on ground conditions and to analyze the data for users mainly in the Department of the Interior and the Department of Agriculture. After placing the first Landsat in orbit in July 1972, NASA, between 1974 and 1978, undertook its largest project: a Landsat for the Department of Agriculture and the National Oceanic and Atmospheric Administration (NOAA) that would examine the impact of weather on crops. As historian Pamela Mack has argued, "Landsat did not represent research that could only be done by the Federal government, and it only marginally fit the definition of a project too risky or too long-term for private investment." That was also the thinking of the Carter administration.[2]

The privatization of Landsat started under President Jimmy Carter. In October 1978, he requested that NASA and the Department of Commerce investigate ways of encouraging private industry to participate in civilian remote sensing, including Landsat, weather satellites, and ocean-observing satellites. It soon became clear that privatizing Landsat would not be an easy task.[3] In March 1983, President Reagan announced the transfer of Landsat, weather satellites, and future ocean-observing satellites to private industry. The Cabinet Council on Commerce and Trade (CCCT) admitted difficulties in privatizing Landsat because, they believed, government was "inherently" unsuited for marketing products and services. The CCCT also urged "some form of Government-assured market for a time," such as a guaranteed minimum purchase agreement, to get the business going.[4]

Eosat, for Earth Observation Satellite Company, a joint venture of Hughes and RCA (now known as the Space Imaging Corporation), won a competitive bid run by the Department of Commerce to purchase Landsat. Eosat started out in a financially weak position. The government provided only a small subsidy for the transition period and guaranteed no data purchases, and Eosat itself never invested significant resources in the system. The Landsat privatization was doomed. The Land Remote Sensing Policy Act of 1992 ended the experiment in privatizing Landsat and transferred oversight from the Department of Commerce to NASA and the Department of Defense. After profitably using Landsat during Desert Storm, the Department of Defense transferred its share of the system to NASA, which launched a sixth Landsat satellite in 1992 (lost in October 1993) and a seventh one in 1999.[5]

NASA would carry out a far more extensive space commercialization program than either the Department of Commerce or the Department of Transportation. The agency was not new to space business, having provided launch services for hire in 1962 to the privately owned AT&T commercial communication satellite *Telstar* for $10 million.[6] It was not until 1972, however, that NASA formed "partnership" arrangements (known as joint endeavor agreements) with private firms to commercialize space. The agency expanded that policy on June 25, 1979, with guidelines that included an offer to "provide flight time on the Shuttle, provide technical advice, consultation, data, equipment, and facilities" and to "enter into joint research and demonstration programs where each party funds its own participation."[7]

In October 1984, following Reagan's National Policy on the Commercial Use of Space, NASA released its own Commercial Space Policy aimed at "the fostering of commercial enterprises in space." NASA echoed the language of the Reagan Revolution and the conservative agenda for space. For instance, the agency declared that barriers to space commercialization, whether "natural" or bureaucratic, "need to be and can be lessened or removed through joint actions by the Government and private enterprise." NASA and industry would cooperate in order "to expedite the expansion of self-sustaining, profit-earning, tax-paying, jobs-providing commercial space activities." Among its initiatives to stimulate research and development in space was the contribution of "seed money" as well as hardware and facilities "at reduced prices." "Seed money" went to both industry and academia through three programs: 1) Academic Commercial Grants, to encourage academic institutions to develop commercial space endeavors; 2) Space Commerce Announcements of Opportunity, to solicit proposals concerning space commerce ventures; and 3) Industry Space Commercialization Grants, to encourage firms to undertake space commerce ventures by reducing their up-front costs.[8]

Of course, NASA would offer rides on its shuttle, too. The agency pledged to make the shuttle more convenient and cheaper for commercial users, who would receive "reduced rates." The agency even announced Get-Away Special flights. Entrepreneurs could launch a shuttle payload weighing less than 200 pounds (91 kg) and taking up less than 5 cubic feet (under 1.5 cubic m) of cargo space for $10,000, which was less than half the 1980 price of $21,000 for the same weight and space. At the 1980 price, NASA already had 200 customers who had put up $500 in earnest money to launch 315 payloads. This was beyond the capacity of the shuttle, which had yet to fly. NASA also promised to integrate and fly "standard" commercial payloads no later than six months after the start of the integration process. It also would reserve space for commercial payloads on the orbiter middeck on each shuttle mission up to twenty weeks before launch.[9]

Charged with carrying out these projects was the Office of Commercial Programs. Established in September 1984, and headed by the assistant administrator for commercial programs, long-time NASA employee Isaac T. Gillam IV, the Office of Commercial Programs also fostered the creation of Industry-University-Government Advanced Research Institutes, which came to be known as Centers for the Commercial Development of Space in 1985. The centers would conduct basic research in partnership with industry and academia mainly in the field of materials sciences, especially electronic materials, metallic alloys, glasses and ceramics, biological materials, and fluid dynamics. Experiments developed through the program would fly on the shuttle or on sounding rockets. NASA originally planned to underwrite three to six such centers at $500,000 to $1 million each; however, in late 1993, NASA abruptly terminated six of the seventeen centers as failures.[10]

NASA also offered entrepreneurs room on space station *Freedom.* Reagan believed that the station would "permit quantum leaps in our research in science, communications, in metals, and in lifesaving medicines which could be manufactured only in space"[11] He supported the space station, a large, expensive government program, the kind of program Reagan decried, because it would promote the commercialization of space. Before 1983, NASA could not persuade policymakers that a space station served any pressing national need. As governor of California, Reagan had made speeches praising the economic and technological spin-offs from space exploration. President Reagan first learned of the space station project during an August 1983 space commercialization conference. A number of business leaders there told him that they needed the facility to develop new products and to create new industries, but said they could not afford to build one on their own. Reagan also received a letter from congressional supporters claiming that a station was "particularly compatible with [his] economic program" and that "without government backing in this largely uncharted area, space development will be unnecessarily delayed."[12]

Once President Reagan bought the station as a space business program, NASA began to study ways to stimulate commercial interest in it. The agency had several consulting firms determine the number and types of businesses interested in conducting operations on the station. One study outlined 200 potential space products and services in communications, materials, and energy development, while another study found electronics companies interested in semiconductor crystal growth research; an electrical company interested in chemical separation studies; metals companies interested in researching castings and solidification; and chemical companies interested in membranes, films, and coatings.[13] The space station, however, failed to materialize, like other efforts to exploit space for profit.

The Wild Frontier

The objective of Reagan space policy was to stimulate entrepreneurship in space, not to create government programs like the space station. Companies had been exploiting space for long-distance communications since the 1950s without government incentives. Even before the launch of *Sputnik*, researchers at ITT's Federal Telecommunications Laboratories and at Collins Radio independently attempted to use the Moon as a natural communication relay satellite.[14] *Echo, Telstar, Relay, Syncom 2,* and *Intelsat I* (alias *Early Bird*) were the first steps toward a global space-based system of telecommunications operating within the framework of Intelsat (International Telecommunications Satellite organization), created in 1964.[15]

By 1980, satellite communications represented a significant and financially rewarding space business, but it was the only business conducted in space. The imminent availability of the shuttle as an inexpensive and regular "highway to space" stimulated a wide range of speculation about, as well as investment in, new ways of commercially exploiting space. This excitement even extended to the popular cinema. The 1982 movie comedy *Airplane II: The Sequel* portrayed a commercial space shuttle company (Pan Universe) that attempted to keep a pilot from revealing the technical problems of the single-stage-to-orbit *Mayflower* shuttlecraft.

While the policy, regulatory, and tax climates cultivated by the Reagan administration favored space business, the shuttle's promise of cheap, reliable space transport further promoted excitement about doing business in space. Indeed, many of the space ventures proposed in the 1970s and early 1980s took the shuttle as their starting point. One of the most successful of the new ventures built around the shuttle was Orbital Sciences Corporation, initially located in Vienna, Virginia. The firm's founders were three graduates of the Harvard Business School: David W. Thompson, Bruce W. Ferguson, and Scott L. Webster. They obtained their initial capitalization in September 1982 from two Texas businessmen, Fred C. Alcorn, president of Alcorn Oil and Gas Co., and Sam Dunnam, president of Centex Communications, Inc. Orbital's first contract was to develop the transfer orbit stage and the apogee maneuvering system for NASA's shuttle.[16]

Among the unsuccessful ventures hitched to the shuttle's star was the Space Transportation Company (SpaceTran) of Princeton, New Jersey, established in February 1979 and headed by Klaus Heiss. It proposed to purchase a shuttle and dedicate it to commercial and foreign users.[17] In 1978, William Good, a Braniff Airlines pilot, formed Earth Space Transport Systems, Inc., to push for legislative and administrative initiatives to privatize a NASA shuttle.[18] The Space Shuttle of America Corporation, a subsidiary of Astrotech International, was prepared to pay NASA $2 billion to purchase an existing shuttle and to start production of a fifth orbiter.[19]

The goal of dedicating a NASA shuttle to commercial payloads reflected a larger enthusiasm among established corporations for doing business in space with the shuttle as the preferred carrier because it promised reliability, frequency, and economy. An example of this enthusiasm was the plan of the Fairchild Space Company, Germantown, Maryland, to put two Leasecraft satellites into orbit in a joint effort with NASA. Fairchild would charge companies and government agencies $2 million to $4 million per month to rent space on a reusable Leasecraft, while NASA would provide launch and satellite retrieval services gratis in exchange for a free ride on a Leasecraft. Although Fairchild predicted that the venture would turn a profit within two years, the enterprise never got off the ground, literally, because of an inability to obtain insurance coverage, a bad omen for those seeking to exploit space for profit.[20]

Another bad omen was the difficulty of Sparx Corporation's search for a launch provider. Sparx was a joint venture between the German aerospace company Messerschmitt-Bölkow-Blohm (MBB) and Stenbeck Reassurance Group to offer commercial remote-sensing services. NASA scheduled the Sparx payload for shuttle flight 13 (manifest no. 41-G). However, because Sparx was unwilling to comply with a list of terms and conditions imposed by NASA, including a guarantee that the firm would not give customers proprietary or de facto proprietary rights to data, NASA refused them the ride. Sparx then turned to Europe for its launcher.[21]

Despite these problems, many large established firms were thinking about doing business in space. Among them were Texas Instruments, Du Pont, Monsanto, Exxon, US Steel, John Deere, and 3M. Growing gallium arsenide and other semiconductor crystals worth $50,000 per pound in the low-gravity environment of space promised purer materials in larger quantities. Microgravity Research Associates, Inc., Coral Gables, Florida, and Microgravity Technologies, Inc., Solana Beach, California, formed to exploit these and other crystal-growing opportunities.[22]

One of the most successful space manufacturing ventures was a joint effort of McDonnell Douglas and the Ortho Pharmaceutical Corporation subsidiary of Johnson & Johnson. The two firms developed an apparatus that used a process known as continuous-flow electrophoresis to isolate a hormone in higher and purer concentrations, and in larger batches, than could be achieved under the influence of gravity. The increase in purity was four to five times, and the amount of material that could be processed in a batch was 716 times greater. The process also could separate enzymes, proteins, and certain cells. Within twelve hours of the shuttle orbiter's landing, NASA delivered the hormone samples from the electrophoresis apparatus to McDonnell Douglas in St. Louis, Missouri, where a project team processed it into 10-milliliter amounts, then delivered it frozen to Ortho Pharmaceutical. By 1984, NASA had flown the electrophoresis apparatus on four shuttle flights, and a fifth was in the works.[23]

Riding high on the wave of this drug success, McDonnell Douglas began talks with three additional pharmaceutical firms about manufacturing other drugs in space using the electrophoresis and similar separation processes. The firm also considered using Fairchild's Leasecraft and a proposed unmanned space "factory." In 1985, NASA and Space Industries, Inc., based in Houston and headed by former NASA engineer Maxime Faget, signed a memorandum of understanding to operate and service the Industrial Space Facility as a joint venture. The shuttle would launch and service the "factory." Wespace, a subsidiary of Westinghouse, and Boeing soon joined the venture. Space Industries greatly overestimated corporate interest in microgravity research and processing, however. It raised only $30 million toward the estimated $700 million cost of the "factory." The Reagan administration pressured NASA to accept the company's proposal that NASA become an anchor tenant, and Congress ordered NASA to set aside $25 million from space station funding as a first installment on the lease. After an independent study of the project by the National Academy of Sciences concluded in 1989 that no pressing need for it existed, support and enthusiasm for the space "factory" dissipated.[24]

Risky Business

Space Industries, Inc., and Fairchild had grounded their hopes on NASA's shuttle. While those companies saw profits coming from businesses in space, other companies hoped to profit from launching those ventures into space. These private launch companies competed with the government-owned shuttle and bet on the technologies of expendable launchers developed earlier by the military-industrial complex for the cold war. In contrast to the recently developed reusable technologies of the shuttle, expendable technologies were older and better understood. Entrepreneurs hoped that by exploiting those older technologies, they could undercut the government prices of the shuttle. Underlying their aspirations was the notion that industry can do anything better than government, a belief consistent with the conservative agenda for space and the Reagan Revolution.

Virtually all entrepreneurs developing their own expendable launchers ultimately were unsuccessful. They found they couldn't raise enough venture capital to stay in business. One example was Project Private Enterprise, founded in 1977. It was going to design and build the *Volksrocket* (the rocket equivalent of the Volkswagen). The *Volksrocket* was a smaller version of the *Sea Dragon*, a liquid-fueled, ocean-landing rocket that Robert Truax, rocket pioneer and founder of Project Private Enterprise, had designed in the early 1960s for the Aerojet General Corporation. Powering the 25-foot-tall (7.6-m) *Volksrocket* were four small LR-101 engines built by the Rocketdyne Corporation and purchased from the Pentagon for only $115 as surplus. The rocket was slated to take off from an unspecified site in Southern California and parachute to a wet landing in the Pacific Ocean. Truax claimed that he would return a profit to investors just by selling the TV rights to the launch of the first manned *Volksrocket.*[25]

Another project failure was one developed by Starstruck, Inc., of Redwood City, California, a company financed largely by Michael Scott, a wealthy cofounder of Apple Computer. Starstruck was developing a prototype of their *Constellation* rocket known as the *Dolphin.* The *Dolphin* was a recoverable booster capable of carrying payloads up to 1,000 pounds (about 450 kg) using a hybrid propulsion system that burned a modified rubber fuel. By 1984, Starstruck had built hardware, tested engines, and made two aborted launch attempts off the California coast.[26]

A third doomed enterprise, which was intended to develop a reusable launcher from available hardware, was TranSpace, Inc. Founded by former Rockwell International aerospace engineer Leonard Cormier in 1967, TranSpace had the initial goal of exploring private support and operation of launchers for profit. In 1981, TranSpace filed with the Securities and Exchange Commission to sell stock in the hope of developing a reusable spaceplane that would launch from the back of a Boeing 747. The three-stage "space truck" would take off from an airport with an unmanned, remote-controlled, recoverable second stage to which was attached piggyback-style a manned third stage. At appropriate altitudes, each stage deployed in turn to put the third stage in orbit. All three reusable stages would land at an airport. TranSpace planned to use the RL-10 engine developed by Pratt & Whitney for the *Centaur* upper-stage booster.[27] The *DC-X* would use the same RL-10 engine.

The one successful start-up firm was Space Services, Inc. (SSI) of Texas. Success came despite an initial, and potentially ruinous, failure. The founder of SSI, Houston real estate developer David Hannah, Jr., met Gary C. Hudson in 1979 at a Space Foundation fund-raiser. *Time* magazine described Hudson as "a college dropout and self-taught engineer." Hudson had been working on various launcher concepts since 1970 and concluded that the main impediment to the commercialization of space was the high cost of transport. He convinced Hannah to build a low-cost expendable rocket using off-the-shelf hardware that would put payloads in orbit at bargain prices. Enthused about the potential profitability of the venture, Hannah raised funds from Texas oil magnates and started SSI, which subcontracted to Hudson as GCH, Inc., to provide the launcher.[28]

Hudson and seventeen engineers (some of whom were former NASA employees) built the *Percheron*, named after a French draft horse, in a Sunnyvale, California, facility. A flatbed truck hauled the *Percheron* liquid-fuel rocket across the country to the SSI launch site on the Gulf Coast at Matagorda, about 50 miles northeast of Corpus Christi. Hannah arranged for $25 million in flight insurance, just in case. The worst *did* happen. On August 5, 1981, during a five-second engine test firing on the launch pad, the *Percheron* exploded, probably as a result of a jammed propellant valve.[29]

SSI persevered. Hannah raised $6 million from fifty-seven investors for a second rocket. Space Vector, Inc., of Northridge, California, built the 37-foot-tall (11.3-m) solid-fuel *Conestoga I*. Space Vector had several years of experience building solid-fuel sounding rockets for scientific and military research. SSI also acquired the expertise of former astronaut Donald K. "Deke" Slayton, who joined as the firm's president. The *Conestoga* successfully completed its first suborbital test flight on September 9, 1982. The flight lasted about 10.5 minutes and the rocket splashed down in the water about 321 miles (517 km) from the Matagorda launch pad.[30]

While SSI and other small start-up firms hoped to create new launchers from extant hardware, the large corporations that had created launchers for the cold war hoped to shift them to the commercial launch market. The conscription of cold war hardware into commercial service proved to be more profitable and far easier than developing new launchers. After all, the cold war had paid the costs of both development and testing, and use had established their reliability.

One of the earliest efforts at recruitment of cold war missiles for commercial launches was made by the Boeing Aerospace Corporation, manufacturer of the *Minuteman* missile. Boeing offered to buy back outdated *Minuteman* missiles from the air force and intended to use them to launch satellites for hire in 1974. The project failed because of objections from the State Department.[31] However, in 1983, NASA agreed to turn over the *Delta* rocket program to Transpace Carriers, Inc., of Greenbelt, Maryland. McDonnell Douglas would build the *Delta* rockets, and Transpace Carriers would sell them as commercial launchers. Also in 1983, NASA transferred the Atlas-Centaur program to the Convair Division of General Dynamics of San Diego, California.[32]

Fig. 1. Despite its catastrophic start, the *Percheron* was the first completely commercial launch vehicle. Funded and built by private enterprise, it took off from a launch site built and owned by SSI. The next commercial launch attempt succeeded and challenged the regulatory environment. (Courtesy of the NASA History Office.)

The Fight for the Sky

Before these launch companies could turn their cold war weapons into plowshares and cultivate customers, they had to confront competition from both the shuttle and Arianespace. The shuttle was a formidable government-subsidized competitor that stymied development of a commercial launch industry in the United States. Between November 1982 and January 1986, it carried twenty-four communication satellites into orbit on eleven flights. Five were for private corporations: *Westar 6*, two *Telstar*s and two *SATCOM*s. Other satellites were for foreign clients, including Canada (four *Anik*s), Australia (two *AUSSAT*s), Indonesia (two *Palapa*s), India (*INSAT*), and Saudi Arabia (*ARABSAT*).[33]

The other competitor, Arianespace, was "a real threat" to the emergence of a commercial launch industry in the United States. As Newt Gingrich warned presciently, "the United States may lose a lucrative market."[34] Arianespace was a private stock company created in March 1980 by European aerospace firms, banks, and the French space agency. The company took over operation of the multinational European Space Agency's *Ariane* rocket, including managing and financing *Ariane* production, organizing worldwide marketing of launch services, and managing launch operations at Kourou, French Guyana. The first full commercial *Ariane* mission under Arianespace control was the rocket's ninth flight in May 1984, when an *Ariane I* successfully lifted the U.S. GTE *Spacenet 1* satellite into orbit. By the spring of 1985, Arianespace held firm orders for orbiting thirty satellites and had options for launching twelve more, representing a combined order book value of about $750 million. Of those orders, half were from satellite customers outside the European home market. Arianespace marketing combined the best of both worlds: the marketing freedom of a private company plus the direct support of government agencies.[35]

Arianespace consistently priced launches below the shuttle. It was able to offer lower prices for a variety of reasons. One had to do with the way it placed satellites in geosynchronous orbit. The *Ariane* rocket lifted satellites into a different orbit, but at the required altitude (23,000 miles; 37,000 km). Small thrusters moved the satellite to the desired orbit. This maneuver required less satellite fuel, and the fuel saving could add a year to the satellite's life expectancy of seven to eight years. The added life might mean a saving of as much as $15 million. In contrast, after the shuttle discharged a satellite, its kick motor had to both raise it to 23,000 miles (37,000 km) and move it into geosynchronous orbit. This double maneuvering required a great deal of fuel. In one case, it also wrecked havoc with the launch insurance industry. A misfiring of the kick motors on a 1984 launch sent a Western Union and an Indonesian satellite into useless orbits. As a result, the cost of insurance skyrocketed. Insurers paid claims of $200 million on the two satellites, more than all the premiums collected over the fifteen-year history of launch insurance.[36]

Such incidents alone did not account for Arianespace's near monopoly of the commercial launch market. *Ariane* rockets also experienced failures, and insurance premiums, which customarily reflect risk levels, were higher for the *Ariane* launches than for the shuttle. More than any other factor, it was Arianespace's lower prices that gave it an edge over the shuttle, or any other potential competitor. Nominally, though, NASA charged less. For the period 1986–1988, NASA announced it was charging $17.5 million to launch a standard-sized satellite into low orbit. However, customers then had to pay an extra $7.5 million for kick motors to raise and move the satellite into a higher orbit, making the total cost $25 million. Arianespace, in contrast, charged $24 million for the same service.[37]

Faced with the devastating competition from Arianespace and the shuttle, Transpace Carriers, which provided McDonnell Douglas *Delta* launchers to paying customers, decided to attempt to block the Europeans from conducting trade in the United States by charging Arianespace with unfair subsidy pricing. Transpace Carriers claimed it could not launch a satellite for less than $45 million, the same service for which Arianespace charged $24 million. Arianespace had a two-tier pricing policy, charging higher prices to the European Space Agency (ESA) and its member states. The French space agency subsidized Arianespace launch and range services, as well as administrative and technical personnel, and ESA member states subsidized Arianespace insurance rates.

On May 25, 1984, Transpace Carriers filed a petition with the Office of the United States Trade Representative (USTR) under Section 301 of the Trade Act of 1974. Transpace alleged that Arianespace engaged in predatory pricing and other unfair trade practices in the sale of *Ariane* launch services. The USTR accepted the case on July 9, 1984, and meetings between European and U.S. government officials began in November. Talks soon turned to comparisons of the subsidies and pricing policies of the shuttle and *Ariane.* On July 17, 1985, President Reagan ruled that the pricing and subsidy practices of Arianespace were neither unreasonable nor a restriction on U.S. commerce because Arianespace practices were not sufficiently different from those of the shuttle to warrant action under the Trade Act of 1974.[38]

One Size No Longer Fits All

Private launch firms had to compete not only with Arianespace but also with the shuttle. The competition was too much for the nascent commercial launch industry. As a result, the Department of Transportation, authorized to license commercial launches under the Commercial Space Launch Act of 1984, did not issue its first license until 1989. The only U.S. response to Arianespace was the shuttle, and as long as the shuttle carried commercial payloads, newcomers found themselves between the Scylla (shuttle) and Charybdis (Arianespace) of competition.

Meanwhile, the shuttle was becoming a policy problem. It was supposed to be a frequently flying, low-cost space "truck," but its operating costs were anything but low. A 1982 General Accounting Office study found that the cost per payload pound, a common measure of launch cost, had been estimated by NASA to be $100, but by one accounting was as high as $1,700.[39] One of the keys to keeping shuttle operating costs low was to fly often so that fixed costs could be spread out over a large number of flights. NASA hoped to fly the shuttle forty-eight times a year, but lowered this schedule to twenty-five flights. Because of the long maintenance and preparation time required between flights, the shuttles have flown no more than nine times per year. During its first twelve years of operation (1981-1992), the shuttle flew fewer than four times a year in five of those years (1981-1982 and 1986-1988).[40]

Shuttle commercial payload pricing became a topic of discussion at the White House in June 1984, as the Transpace Carriers petition with the Office of the USTR against Arianespace was under consideration. Debate centered on whether prices should be raised to allow private enterprise to compete better with the shuttle. NASA claimed that shuttle pricing was not the issue; customers wanted quality and reliability (the shuttle) more than low pricing (Arianespace). In July 1984, for example, NASA Administrator James M. Beggs argued that Transpace Carriers had "yet to succeed in winning a launch contract," even though they underbid the shuttle, because "customers are basing their decisions on quality and reliability of service."[41]

Shuttle prices could not be raised, Bud McFarlane, assistant to the president for national security affairs, wrote to Secretary of Transportation Elizabeth Dole, because Arianespace would be the primary beneficiary. Moreover, banning commercial payloads from the shuttle would have serious foreign policy implications. The shuttle was "a significant and highly visible instrument of our foreign policy . . . an effective means for promoting international cooperation, good will and technological growth among our friends and allies."[42]

The Reagan administration was not just at an impasse on shuttle pricing. It had tried to foster a U.S. commercial launch industry while allowing the shuttle to compete against Arianespace for commercial payloads. This fatal policy flaw revealed itself on January 28, 1986. The tragic loss of the *Challenger* and its crew was a watershed moment for the U.S. space program, for NASA, for the Department of Defense, and for space commerce. What made the accident so damaging, aside from the loss of human life, was the policy that placed all

NASA and military payloads aboard the shuttle. The dependence on the shuttle as the nation's "primary" launch system now impaired the ability of the nation's defense and intelligence agencies to place payloads into orbit. The nation needed a variety of launchers, both reusable and expendable. NASA in particular needed to relinquish its dependence on the shuttle. The loss of the shuttle had an even deeper impact on the U.S. commercial space industry. NASA put on hold companies with payloads scheduled for future shuttle flights. The shuttle would not fly again until the cause of the *Challenger* accident was known and the shuttle was made safer and more reliable.

National Security Decision Directive 254 (NSDD 254), released shortly after the *Challenger* accident, took NASA and the shuttle out of competition with potential commercial launch providers. Specifically, it stipulated that "NASA shall no longer provide launch services for commercial and foreign payloads subject to exceptions for payloads that: (1) are Shuttle-unique; or (2) have national security or foreign policy implications."[43] By "Shuttle-unique," the directive meant payloads requiring either human intervention or facilities available only on the shuttle. NSDD 254 helped to create a commercial launch industry in the United States by clearing away shuttle competition. At the same time, though, it gave Arianespace a monopoly of the market—at least for the moment.

Subsequently, NSDD 254 became part of a fuller revised national space policy approved by President Reagan on January 5, 1988, but released in February 1988, after a lengthy review of existing space policy ordered by the president. The 1988 national space policy soon made its way into legislation as the Commercial Space Launch Act Amendments of 1988.[44] The main objective of the new policy was to promote the commercial launch industry. The 1988 national space policy overthrew the long-standing notion of the shuttle as the nation's "primary" launch system. "The United States national space transportation capability," the new space policy declared, "will be based on a mix of vehicles, consisting of the Space Transportation System (STS), unmanned launch vehicles (ULVs), and in-space transportation systems." Essentially, NASA would use the (partially) reusable shuttle, and the Department of Defense would rely on expendable launchers. The policy specifically excluded NASA from maintaining its own expendable launchers. If the agency wanted to use an expendable launcher, it would have to turn to the Pentagon or industry. The policy of assigning expendable launchers to the Pentagon and reusable launchers to NASA would remain in effect into the next decade and beyond, buttressed by intervening space policy declarations, despite three changes of leadership within the White House.

The 1988 space policy also directed both NASA and the Department of Defense to "utilize commercially available goods and services to the fullest extent feasible, and avoid actions that may preclude or deter commercial space sector activities except as required by national security or public safety." Further measures to "encourage, to the maximum extent feasible, a domestic commercial launch industry" sought to facilitate the use of government ranges and facilities by commercial launch operators. The 1988 national space policy was consistent with the Reagan administration's stand against subsidizing the commercial launch industry. "The United States Government will not subsidize the commercialization of ULVs [unmanned launch vehicles], but will price the use of its facilities, equipment, and services with the goal of encouraging viable commercial ULV activities in accordance with the Commercial Space Launch Act." Instead of direct subsidies, then, the Reagan administration would stimulate the creation of a commercial launch industry by a market mechanism, the prices paid by commercial launch providers for government goods and services (indirect subsidy). Despite these government interventions in the market, the 1988 national space policy stipulated that: "The United States Government will encourage free market competition within the United States private sector."[45]

The Frontier Opens

Removing the shuttle from the commercial launch market was the final step in stimulating the creation of a private launch industry. Given the political values of the Reagan administration, it was illogical for the government to be in the business of selling rides into space, especially when this practice hindered private companies from competing. Now companies could conduct business in space, whether telecommunications, remote sensing, or manufacturing, and companies could provide the ride into space.

The full impact of NSDD 254 became apparent in 1987, when established aerospace giants signed contracts to launch satellites on rockets developed for the cold war. In August 1987, for example, Martin Marietta, manufacturer of the *Titan,* signed a $219 million contract with Intelsat to launch two satellites. McDonnell Douglas had contracts with India, the International Maritime Satellite Organization (INMARSAT), and Hughes Aircraft (for two British direct broadcast satellites, *BSB-R1* and *BSB-R2*) for launches on its *Delta* rocket. The new launch market favored these larger established corporations over start-up firms. Space Services, Inc., announced several launch reservations for its *Conestoga* rocket. American Rocket Company (AmRoc) of California; Conatec, Inc., in Maryland; and E′ Prime Aerospace Corporation, Titusville, Florida, all applied for licenses to launch various sounding rockets. Also, Orbital Sciences Corporation, in partnership with Hercules Aerospace Corporation, soon began offering its two-stage *Pegasus* vehicle to commercial customers.[46]

The real breakthrough year, though, was 1989. As the OCST reported to Congress: "Fiscal Year 1989 was a turning point for the Office of Commercial Space Transportation (OCST) and the US commercial launch industry. . . . [and] the beginning of a new era in the history of US space endeavors."[47] The new era was evident in the amount of launch business that the OCST licensed. In 1989, three years after the *Challenger* accident, the OCST issued four commercial launch licenses. Five more license applications were pending. The first licensed launch took place on March 29, 1989.[48] An SSI *Conestoga* lofted a microgravity research platform, marking the inauguration of SSI's microgravity business, called Consort. It was also a score for NASA's efforts to foster space commerce. Consort had originated in 1988 in NASA's Office of Commercial Programs and the Consortium for Materials Development in Space (CMDS) of the University of Alabama at Huntsville, which managed the program. The CMDS was one of seventeen NASA Centers for the Commercial Development of Space.[49]

The real story, however, began with the second licensed launch. On August 27, 1989, a McDonnell Douglas *Delta* rocket lifted off from Cape Canaveral carrying a British Satellite Broadcasting Company communication satellite.[50] The commercial launch industry would belong to three large rocket manufacturers: McDonnell Douglas *(Delta),* General Dynamics *(Atlas),* and Martin Marietta *(Titan).* Typical was the success of General Dynamics. In 1989, the firm had contracts in hand to launch a dozen satellites for U.S. and foreign agencies and one business customer. That number included three satellites for the Department of Commerce *(GOES),* one for the European Telecommunications Satellite Organization (Eutelsat), three for Intelsat, one for the U.S. Navy, one for NASA and the U.S. Air Force jointly *(CRRES),* two for the Orion Satellite Corporation, and the *SAX* scientific satellite for the Italian Space Agency (normally an Arianespace payload).[51]

The smaller start-up launch firms fared less well. They were smaller because they were new, and they lacked the established reputations, contacts, and capital resources of the big three aerospace corporations. Most importantly, though, the government had paid for the development of the launch systems of the big three in the name of the cold war. SSI, one of only two small firms that survived, found itself in troubled financial waters in 1990, when its chief financial backer pulled out.[52] EER Systems Corporation of Vienna, Virginia, purchased SSI, along with its Consort microgravity business, its *Conestoga* technology and launch pads, and the accumulated experience and knowledge of its managers, including Deke Slayton.[53] On October 23, 1995, the latest *Conestoga* rocket design broke up off the Virginia coast. That single launch failure marked the end of the *Conestoga*, EER Systems, and the COMET program, whose payload was to be launched.[54] Orbital Sciences Corporation fared much better, despite repeated launch failures. It continued to offer *Pegasus* and other small launchers to paying governmental and corporate customers. In fact, of all the small start-up commercial launch firms in existence in 1989, Orbital was the only one still in business in 2002.

Despite unleashing the three giant aerospace firms on the global launch market, the United States was still in second place behind Arianespace.[55] However, the U.S. commercial launch industry was now a contender in the global market, and that was the goal of space policy under the Reagan administration. President Reagan had raised space policy to the same status as other national policy issues. This was a radical departure from earlier policy. Moreover, it was nothing less than the extension of the Reagan Revolution into space, and the achievement of the conservative space agenda. That policy, though, failed to help start-up firms to develop new launch systems for the commercial market. A more active and interventionist government policy would have been needed to achieve that goal. Instead, the conservative economic approach of the Reagan administration favored the expendable *Titan, Delta,* and *Atlas* launchers developed by the military-industrial complex for the cold war and operated commercially by the same aerospace firms. The commercialization of space thus became a new arena for the use of cold war hardware by the military-industrial complex. This was a fortuitous trend. The first licensed commercial launches took place in 1989, as the cold war was coming to a close.

CHAPTER 3

Space Warriors

The commercialization of space was only half of the Reagan Revolution in space, half of the conservative space agenda. The other half was the militarization of space. The Reagan administration did more than advance the militarization of space, however. It took the cold war in a new direction that, in many ways, repudiated the conduct of the cold war initiated by President Nixon that persisted throughout the 1970s. The two cardinal facets of Nixon's cold war policy rejected by Reagan were détente and the 1972 Antiballistic Missile Treaty (SALT I). Instead, President Reagan chose to heighten the struggle against what he termed "the evil empire."

The eight years of the Reagan presidency were a period of strained relations between the U.S. and the USSR. The Soviet invasion of Afghanistan in December 1979 destroyed what was left of the era of détente. The boycotts of the 1980 and 1984 Olympics showed the strain, as did the shooting down of Korean Air Lines flight KAL 007 in September 1983. President Reagan denounced that act as "wanton, calculated, deliberate murder." His rhetoric reflected the diplomatic tension with references to the Soviet Union as "the focus of evil in the modern world." That was just the beginning.

Reagan expressed disdain for and distrust of the SALT I agreement, which he and others believed the Soviet Union had violated. He also spoke against the reigning defense philosophy known as MAD for mutual assured destruction. The MAD strategy, simply stated, was that each party would be able to wreak destruction on the other, even if an initial strike substantially reduced the missile and nuclear forces of one side. Each side essentially became the hostage of the other. Reagan's stance against MAD combined with ongoing studies of high-energy lasers and satellite weaponry to become the Strategic Defense Initiative (SDI). The agency charged with erecting SDI, the Strategic Defense Initiative Organization (SDIO), became the institutional home of the Single Stage To Orbit (SSTO) program and the *DC-X*.

Over the course of the cold war, the military use of space was a constant. Starting in the 1960s, the U.S. military relied on satellites for photographic, electronic, and oceanic reconnaissance, early warning of offensive missile launches, detection of nuclear explosions, communication, navigation, weather, and geodetic information. *Sputnik* marked a milestone in the militarization of space. One might even say that, at least at one point in the conflict, all space efforts served a cold war end. The Apollo program *was* the cold war at one level. While fighting the cold war, the United States simultaneously hoped to assuage the conflict through treaty negotiations. The construction of defensive systems served as a bargaining chip in treaty negotiations, while treaties helped to limit the seemingly boundless search for, and construction of, new weapons. The Strategic Defense Initiative was no exception.

SDI took military space policy in a new direction by proposing to place defensive weapons in space. Previous space strategies had positioned defensive weapons on the ground until their use in space to "kill" enemy satellites. The first ground-based antisatellite (ASAT) systems involved launching a killer satellite atop a booster so that its orbit matched that of the target, tracking the satellite target, and detonating the killer satellite near the target. ASAT is not a designation for a single weapon system but rather a generic term covering anything that can be used to attack, disable, or destroy a satellite from Earth or (in the case of SDI) from space.

The 1972 ABM Treaty

The cold war of the 1950s was global and total. While the Berlin blockade and other conflicts heated up in Europe, war erupted in Korea, a war that involved the third party to the cold war, China. *Sputnik* truly had a profound impact on the prosecution of the cold war because it demonstrated the Soviet Union's ability to hurl nuclear bombs at the United States. In response, among many other measures, the United States and its NATO allies began negotiations to install missiles in Europe closer to the Soviet Union (which led in turn to the Soviet Union's placement of missiles in Cuba).

Also in response to *Sputnik*, the Eisenhower administration assembled the nation's nonmilitary space programs and centers into a single civilian agency, the National Aeronautics and Space Administration (NASA). Although Eisenhower's intent had been to create a space agency independent of the defense establishment, his successor made NASA serve the cold war. On May 25, 1961, President Kennedy initiated the space race, a new cold war battleground. In his speech, Kennedy referred to the flight of cosmonaut Yuri Gagarin as "the dramatic achievements in space which occurred in recent weeks." The United States would have to challenge those space feats, "if we are to win the battle that is now going on around the world between freedom and tyranny," the president warned.[1] The country perceived that it was in a space race with the Soviet Union, and the United States appeared to be losing.

Following the president's speech drafting the space agency into the service of the cold war, NASA undertook Kennedy's challenge to put an American astronaut on the Moon by the end of the decade. The cold war and the nation's space programs now were joined. This was truly "total" warfare that conscripted even civilians and civilian agencies into a global struggle. By 1972, however, the space race was over. America had won. In December 1972, *Apollo 17* astronauts Eugene A. Cernan (spacecraft commander), Harrison H. "Jack" Schmitt (lunar module pilot), and Ronald E. Evans (command module pilot) became the last people to visit the Moon's surface. The Apollo program was over. Instead of competing in space, NASA now entered an era of comparative cooperation or "détente" with the Soviet Union. The most illustrious example of that cooperation was the Apollo-Soyuz Test Project (ASTP). In July 1975, a *Soyuz* and an *Apollo* spacecraft met and docked in Earth orbit. The crews conducted a number of experiments, then returned to Earth.[2]

The detachment of the civilian space program from the cold war effected during the Nixon administration had tremendous immediate and long-term consequences for civilian uses of space. The goal of landing an American astronaut on the Moon was but a single effort focused on achieving one great national policy goal. That dramatic human effort often eclipsed the science undertaken via remote sensing instruments aboard *Pioneer 4*, and the *Ranger, Surveyor,* and *Lunar Orbiter* series of probes, not to mention the *Mariner* probes sent to Venus (1962 and 1967) and to Mars (1964-1965 and twice in 1969) during the same decade. The period from 1971 to 1978, though, in the words of NASA's then-director of planetary programs, was a golden age of planetary exploration.[3] The planets had replaced the Moon as the major focus of the space program. *Mariner* flights visited Mercury and returned to Mars and Venus, while *Pioneer 10* and *Pioneer 11* flew by Jupiter and Saturn before passing outside the heliosphere. In 1976, *Viking* landed on Mars, and in 1978, *Pioneer Venus 1* and *2* created radar images of the planet's hidden surface and studied its atmosphere, ionosphere, and magnetosphere. Between the *Viking* and the *Pioneer Venus* probes, NASA launched the two *Voyager* spacecraft in

1977 on their grand tour of the outer planets Jupiter, Saturn, Uranus, and Neptune.

The advent of the shuttle in the 1980s (though initiated by President Nixon) once again changed the nature of the space program by facilitating a growing inward-looking, Earth-centered tendency that bloomed during the 1980s. It also kept human space flight alive. Prior to 1980, NASA's space program consisted mainly of lunar and planetary exploration and the design and construction of spacecraft and launchers to achieve those missions. Planetary exploration did not end, but slowed significantly with only a few grand projects, Galileo and Magellan. The shuttle was ideal for enabling the conservative space agenda, which emphasized exploiting, rather than exploring, space, especially space applications (business and defense) in near-Earth space. In addition, because the shuttle carried military and intelligence payloads, the line between civilian and military missions blurred. Was NASA again in the service of the cold war? The question appeared to be resolved in 1986, when the shuttle stopped carrying defense and commercial payloads.

The removal of the civilian space program from the cold war during the Nixon administration was but a small sign of the easing of conflict between the United States and the Soviet Union. Tensions had mounted continually throughout the 1960s, from the Bay of Pigs (April 1961) to the Berlin Wall (August 1961) to the Cuban Missile Crisis (October 1962) and to Vietnam. Both the United States and the USSR also intervened in Latin America and in newly independent African states. At the same time, though, President Kennedy undertook measures to limit the severity of the cold war, such as the "Hot Line," the Partial Test Ban Treaty, and United Nations Resolution 1884 (XVIII), which formed the basis for the 1967 Outer Space Treaty. That agreement banned weapons of "mass destruction" from all "celestial bodies." It specifically forbid military bases, installations and fortifications, weapons testing, and military maneuvers in outer space.[4]

The cold war was expensive. The nuclear missile arsenals of the United States and the Soviet Union were growing to a colossal magnitude. As each side built more and more offensive weapons, each side felt constrained to build larger and more extensive defensive weapons. And as each side built up its defenses, each felt compelled to build more weapons to overcome those defenses. The pursuit of the cold war at the height of the Vietnam War presumed a robust economy like that which the nation had experienced during much of the 1950s and 1960s. By the end of the 1960s, however, America was in the midst of a major recession, and expenditures on the Vietnam War and investments in education and public welfare were mounting. The resulting economic and budgetary dilemmas pitted one policy priority against another. Why should the country spend on space rather than on the needs of the poorest citizens? One critic, Rep. Ed Koch (D-NY), put the argument in more concrete terms: "I just for the life of me can't see voting for monies to find out whether or not there is some microbe on Mars, when in fact I know there are rats in the Harlem apartments."[5]

The United States and USSR continued to fight the cold war while still trying to negotiate limits to the conflict and to spending on weapons. Nixon's first four years in office was a time of many transitions. It began with the 1968 Tet Offensive, followed by the death of Ho Chi Minh and the start of U.S. troop withdrawals in 1969. Meanwhile, Henry Kissinger had begun secret peace talks. In 1972, the last American ground troops left Vietnam. Nixon's eight-day trip to China in February 1972 suggested that the cold war was taking a new direction. This cold war rapprochement shifted the focus to a competitive, but somewhat more controlled, arms race.

Exemplifying the new direction was the successful negotiation of an arms treaty with the Soviet Union, SALT I. At the same time, though, the country developed an operational antiballistic missile defense system, Safeguard, that served as part of the country's negotiating strategy. Safeguard originated in a study of antiballistic missile options commissioned by President Nixon after his inauguration in January 1969. Safeguard gave first priority to protecting U.S. land-based retaliatory forces. It did not protect submarine forces or the civilian population. After widespread public debate and heated battles in Congress, Safeguard received official approval to begin site development, but only by the narrowest of Senate margins, with the vice president breaking a 50-50 vote.[6]

The Nixon administration then used Safeguard as a bargaining chip in its ongoing Strategic Arms Limitation Talks (SALT) with the Soviet Union on limiting antiballistic missiles and other strategic arms. Although the army completed construction of the Mickelsen Safeguard site near Grand Forks, North Dakota, in October 1975, Congress cut off funding less than a year after the site became operational. The Pentagon dismantled the launchers; command of the large perimeter acquisition radar transferred to the North American Air Defense Command (NORAD).[7]

SALT I produced an interim agreement in 1972 on offensive arms and a treaty restricting the deployment of antiballistic missiles to no more than two sites in either nation, with no more than a hundred single-warhead interceptors per site. The two parties amended this provision in a 1974 protocol to a single site in each country. The ABM (antiballistic missile) Treaty, as it became known, further restricted development, testing, and deployment of antiballistic missile systems or components, not permanently fixed and not land-based. Agreed Statement "D" of the treaty subjected "ABM systems based on other physical principles" to "discussion in accordance with Article XIII [the Standing Consultative Committee] and agreement in accordance with Article XIV [amendments and review] of the Treaty." The ABM Treaty, signed May 26, 1972, clearly excluded the development, testing, and deployment of space–based defensive weapons, such as those proposed later for use in the Strategic Defense Initiative. While both parties agreed that the ABM Treaty would be of "unlimited duration," either side could withdraw from the treaty anytime its "supreme interests" were deemed jeopardized by the treaty's provisions, provided the withdrawing party notified the other side of its intention six months in advance.[8]

Satellite Warfare

The 1972 ABM Treaty did not ban antisatellite (ASAT) systems. Neither side wanted to give up a space weapon that both sides were developing. As early as 1958, design studies of satellite interception systems had been part of Project Defender run by the Pentagon's Advanced Research Projects Agency (ARPA), as well as various classified air force, army, and navy projects. U.S. investment in ASAT development increased considerably in 1968, following the discovery that the Soviet Union had begun testing a satellite interceptor. After five successes out of seven attempts, Soviet ASAT tests suddenly stopped in 1972, probably because of a combination of budgetary, political, and technical factors. The USSR resumed ASAT testing in February 1976.[9]

The resumption of Soviet ASAT tests galvanized the Ford administration into authorizing the development of an ASAT system during the last months of Gerald Ford's incumbency. President Jimmy Carter continued the ASAT project, while seeking to revive existing arms control negotiations. With its decision to pursue an operational ASAT, the United States once again was engaged in a space race, and the Soviet Union was in the lead. On June 18, 1979, Jimmy Carter and Leonid Brezhnev signed the SALT II Treaty in Vienna. By setting limits on offensive, not defensive, missiles,[10] SALT II maintained the status quo on research in laser weapons and antiballistic missile defense systems. The USSR again resumed ASAT testing in April 1980.

The Reagan administration took a tough stand on Soviet ASAT testing. On August 20, 1981, Foreign Affairs Minister Andrei Gromyko presented a draft treaty to the United Nations that proposed to supplement the Outer Space Treaty by precluding weapons not covered by the definition of weapons of mass destruction. The treaty would not include ASATs, however. The Reagan administration dismissed the proposal as a hypocritical propaganda ploy. The United States made no effort to either make a counter proposal or give any indication that it was interested in resuming bilateral talks on the subject. Instead, in October 1981, Secretary of Defense Caspar Weinberger affirmed the nation's intent to develop an operational ASAT system.[11] This was the same stratagem utilized successfully by the Nixon administration, which negotiated the 1972 ABM Treaty while deploying Safeguard.

During the late seventies, increasingly alarming reports of Soviet progress in developing directed energy weapons using lasers and particle beams emerged in the American press. Potentially, the USSR could arm ASATs with such weapons. In May 1977, for instance, *Aviation Week & Space Technology* published an article based on information provided by former Director of Air Force Intelligence Maj. Gen. George J. Keegan. The article claimed that the Soviet Union was in the advanced stages of developing an operational particle-beam weapon.[12] The article was credible, because Soviet scientists were renowned for their laser and particle beam research. The extent of their scientific progress was evident from their contributions to scientific literature and symposia, as well as from regular exchange visits with U.S. scientists.

Space-based laser weapons are an ideal defense against enemy missiles. Although ICBMs travel at very high speeds, 14,000 to 17,000 miles per hour (about 22,500 to 27,000 km per hour), light travels about 50,000 times faster. Therefore, once an incoming missile is detected, launching a missile counterattack takes far more time than a near instantaneous, and equally lethal, counterattack by a laser weapon. The United States was not without its own particle-beam and laser weapon research, which started under ARPA's Project Defender virtually from the agency's creation. Laser weapons received increased interest following the invention of the gas-dynamic laser in the late 1960s.[13]

Not everyone was happy with the limits imposed by the 1972 ABM Treaty.

These discontents believed that the treaty marked a triumph of the MAD strategists. They also believed that the Soviet Union was preparing to fight and win a nuclear war, and that they were investing massive sums of money in civil defenses. Building up American antiballistic missiles defenses, therefore, was all the more essential. In 1976, in order to garner expert support for this position, CIA Director George H. Bush convened a group of "experts" known as Team B (to distinguish it from Team A, the CIA analysts who normally performed this function) to reevaluate intelligence estimates of the Soviet Union's global strategy. Team B members were known for their pessimistic views of Soviet defense planning and came from both within the government (Arms Control and Disarmament Agency, State Department) and outside the government (including two retired generals). According to one of the retired generals, Lt. Gen. Daniel O. Graham, former head of the Defense Intelligence Agency (September 1974 to December 1975), Team B based its reappraisal on two factors. The first was a CIA estimate published in October 1976 that placed Soviet defense spending at 11 to 13 percent of Soviet Gross National Product, instead of the 6 to 8 percent previously estimated. The second factor was the allegedly huge Soviet civilian defense effort.[14]

Star Wars

Laser weapons, antiballistic missile systems, critics of MAD, and opponents of the 1972 ABM Treaty came together in the Strategic Defense Initiative (SDI). The SDI also was a case of science fiction and defense thinking imitating and perhaps influencing each other. In 1977, a science fiction film featuring state-of-the-art special effects drew Americans to movie theaters in unprecedented numbers. It would have a profound impact on antiballistic missile defense thinking. The film, *Star Wars,* portrayed a struggle between an evil empire and a rebel alliance. The rebels' plan was to destroy the evil empire's new death star, a space-based battle station with the power to destroy an entire planet. Meanwhile, in our own galaxy, designs and technologies for orbiting battle stations were under development in the military laboratories of the United States and the Soviet Union. The popular media played on the parallel between the *Star Wars* death star and actual orbiting battle stations. The film also appeared to give credence to the proposed orbiting battle stations. Beam weapons and space warfare more and more took on the label "Star Wars" in both the media and popular language.

Following the release of *Star Wars,* Clarence A. Robinson, Jr.,[15] who had been covering antiballistic missiles and directed energy weapons for several years, described an antiballistic missile defense system that used space-based battle stations armed with directed energy weapons in the October 16, 1978, issue of *Aviation Week.* According to historian Donald Baucom, the article may have been the first appearance of the space-based battle station concept in the open literature.[16] The system's feasibility was based on low-cost shuttle flights into orbit. Here again was the shuttle promising to empower the conservative space agenda.

Robinson carefully avoided naming the source of his information. He described the individual as "an anonymous official" of the Lockheed Corporation and noted only that he was quoting extensively the words of an "industry official connected with laser weapons work" and "a Lockheed study of lasers." The anonymous official was Max Hunter. The industry study was Hunter's "Strategic Dynamics and Space-Laser Weaponry," his so-called Halloween paper, completed on October 31, 1977.[17]

Hunter would become a key figure in promoting what became the Strategic Defense Initiative. He started working on space-based lasers at Lockheed early in the decade while running the company's entry in NASA's space shuttle program. The low-cost launcher seemed perfect for orbiting laser battle stations. In the belief that their future antiballistic missile business would be quite large, Lockheed formed a corporate-wide committee, with Hunter in charge, to consider all kinds of antiballistic missile systems. About this time, Hunter also learned about the new high-power gas dynamic lasers from Joseph Miller of TRW.

A team of scientists at AVCO Everett Research Laboratory created the first gas dynamic laser in 1967. Gas dynamic lasers combined the conventional carbon dioxide laser with the aerodynamic expansion of hot gas. A dynamic laser excited carbon dioxide gas electrons to a higher energy level, just like a standard laser, then expanded the hot gas through narrow nozzles into a vacuum. The result was power measurable not in a few watts or even a few thousand watts, but laser powers up to 60,000 watts.[18]

In August 1975, Lockheed Missiles and Space Company completed a detailed study titled "Advanced Space-Based Strategic Concept Study." ARPA funded it, in cooperation with the Army Missile Command Redstone Arsenal, Alabama. The study's purpose was to determine the technologies needed to develop an advanced space-based, high-energy laser weapon system. Lockheed Missiles and Space Company subsequently won the competition to pursue both the Talon Gold program, which concentrated on target acquisition and precise laser beam pointing, and the Large Optics Demonstration Experiment, to show precise beam control within the space-based laser weapon platform.[19]

The 1978 *Aviation Week* article was a glimpse into what the future might hold. Hunter, a member of the Gang of Four, soon attempted to persuade Congress to fund a space-based antiballistic missile defense system. The Gang of Four also opposed the MAD strategy. According to one account, the Gang started because H. Alan Pike, then ARPA's deputy director of high-energy laser research, wanted to speed up the level of research. He joined forces with Angelo Codevilla, then a staff member of the Senate Select Committee on Intelligence and an aide to Sen. Malcolm Wallop (R-WY), who served on the intelligence committee. Together, they rounded up the Gang of Four: Max Hunter of Lockheed; Gerald A. Ouellette of Charles Stark Draper Laboratories; Joseph Miller of TRW; and Norbert Schnog of Perkin Elmer. Each was an expert in a particular component technology needed to put together a space-based laser: Miller knew chemical lasers; Schnog, optics; Ouellette, pointing and tracking technologies; and Hunter, system engineering.[20]

The Gang of Four put together a system that they thought could be completed in a relatively short time (four to eight years). It consisted of a constellation of eighteen laser-equipped battle stations assembled in space from components that the shuttle carried there. Organized into three rings, the constellation of battle stations would orbit at an altitude of 1,087 miles (1,749 km) and would be capable of defending U.S. *Minuteman* missiles against an attack by Soviet *SS-18* missiles. In October 1979, the Gang of Four briefed their proposal on Capitol Hill, with Senator Wallop and Angelo Codevilla making the arrangements.[21]

The Gang of Four's campaign elicited an immediate reaction. The Department of Defense frowned on selling defense projects directly to Congress without the concurrence of the military services. In this case, army ballistic missile defense generals perceived the civilian Gang of Four as poaching on their turf. A 1981 study on space-based lasers commissioned by the Senate Armed Services Committee, and undertaken on behalf of the Defense Space Board, advised against the Gang of Four's proposal. Despite these Pentagon efforts to squash the proposal, a group in favor of space-based lasers formed around Senator Wallop.[22] Thus, before President Reagan announced the Strategic Defense Initiative, a group in Congress already supported it.

Another group outside of government with its own plan for a space-based antiballistic missile defense system, and against the MAD strategy, was High Frontier. The man behind High Frontier was its founder, Lt. Gen. Daniel Graham. Graham had advised Ronald Reagan on national security matters during his gubernatorial and presidential campaigns. After Reagan's election, Graham and Gen. Robert Richardson, USAF (Ret.), founded Project High Frontier with help from members of the president's kitchen cabinet and in affiliation with the American Security Council Foundation's Wedemeyer Strategy Center. The center was an IRS 501(c)(3) "education and research organization" that allowed donors to the project to receive a tax benefit. Those donors provided the $250,000 Graham needed to undertake Project High Frontier and to develop a concrete plan of antiballistic missile defense. A schism between Graham and the American Security Council Foundation resulted in the transfer of Project High Frontier to its new home at the Heritage Foundation.[23]

Founded by Edward Feulner and Paul Weyrich in 1973, the Heritage Foundation was (and still is) a conservative think tank. Its major underwriter was beer magnate Joseph Coors. The foundation diffused its brand of conservative ideas through conferences and publications and supported conservative legislation on a variety of issues. It was a vigorous critic of liberalism at home and communism abroad.[24]

Graham thought that by redirecting the arms race to space, where he believed the United States held the technological advantage, the country would achieve a "technological end run" around the Soviets and once again establish U.S. strategic superiority. The High Frontier proposal, completed early in 1982, called for the establishment of the global ballistic missile defense (GBMD). The GBMD used space-based satellites loaded with small interceptor missiles as a forward-based defense. Boeing employees acting on their own provided much of the technical input for the High Frontier study. Graham's High Frontier group believed the GBMD could be deployed more rapidly and for less cost than one using directed energy weapons. In any case, particle-beam weapons could be added to the GBMD system at a later date, as could piloted spaceplanes.[25]

High Frontier aimed its vision at the public at large, not just the White House and Congress. Graham feared that the "federal bureaucracy" would strangle his infant idea in its cradle, and he believed that the Defense Department viewed High Frontier as "a threat to the entire ICBM modernization program." "Small wonder then," he wrote, "that the large bureaucratic complex involved with strategic nuclear offense looked on a strategic non-nuclear defensive concept with hostility."[26] Graham initially figured among the small cabal of men meeting in the offices of Edwin Meese to formulate strategic defense policy in the fall of 1981. However, he soon was on the outside looking in, but still well connected to political circles. As a result, Graham took his High Frontier proposal to the public. Using populist slogans such as "assured survival" and "a defense that defends," High Frontier had as one of its goals the generation of favorable public opinion regarding missile defenses partly, at least, to muster enough political support so that a subsequent Democratic administration could not terminate the project.[27]

The Strategic Defense Initiative

Whatever role the lobbying of Graham, Hunter, and others played, the individual who ultimately made the decision to create a space–based missile defense system was President Ronald Reagan. The country would not take up a large, expensive government program like the Strategic Defense Initiative (nor the space station) without the approval of President Reagan. Reagan was disposed favorably toward antiballistic missile defense and against MAD, as he made clear several times, even as early as his 1976 bid for the Republican nomination. Reagan visited the headquarters of the North American Air Defense Command beneath Cheyenne Mountain in Colorado during the summer of 1979. Following the tour, Reagan asked the commander what would happen if the Soviets were to fire a single missile targeted at a U.S. city. The commander answered that, although the missile would be detected shortly after launch, the United States could not stop it. On the flight home, Reagan and Martin Anderson, a policy adviser to the Reagan campaign, decided it was time to reexamine antiballistic missile defense.[28]

The process that led to Reagan's call for creation of a space-based defense was slow and took many turns over the year and a half between the initial September 1981 meeting in Meese's office and Reagan's so-called "Star Wars" speech. That story has been told in some detail elsewhere.[29] Shortly after his Orlando speech, in which Reagan described the Soviet Union as an evil empire (echoing the *Star Wars* film), the president instructed his national security advisor Bud McFarlane to draft an addendum to his so-called threat speech scheduled for the end of March. The threat speech was Reagan's annual address in support of his defense buildup. It typically painted a grim picture of Soviet militarism and the task of meeting that challenge. Working quickly and secretly, McFarlane began drafting the speech.[30] Top Reagan administration officials learned of its contents only days before the president delivered it on March 23, 1983.

The bulk (about four-fifths) of that address was Reagan's familiar threat speech. Toward the end, Reagan shifted his tone from "threat" to "hope." He announced that defense technologies had advanced to the stage where the United States could hope to prevent nuclear aggression by developing a defense system that would save lives, rather than avenge them. This was a "formidable technical task" that could "take years, even decades, of effort on many fronts." Reagan then called upon the American "scientific community who gave us nuclear weapons to turn their great talents to the cause of mankind and world peace; to give us the means of rendering these nuclear weapons impotent and obsolete." The effort was to be consistent with the terms of the 1972 ABM Treaty and would begin with the establishment of "a comprehensive and intensive effort to define a long-term research and development program" to find a defense against nuclear ballistic missiles.[31]

Reagan's address, dubbed by many critics his "Star Wars" speech, had a great public impact and emotional appeal. The New Right excitedly welcomed it. On January 6, 1984, National Security Decision Directive 119 formally established the Strategic Defense Initiative. As it gradually materialized between 1983 and 1985, SDI became the Pentagon's largest single research and development program.[32] National Security Study Directive 6-83, signed April 18, 1983, commissioned two study groups to assess the technologies and strategic implications of the initiative. The first group, officially called the Defense Technologies Study Team (DTST), took its informal name, the Fletcher panel, from its chairman, James Fletcher. The DTST recommended a major technology development effort to accelerate the advance of missile defense technology. The second group, also named after its chairman, Fred Hoffman, bore the official name Future Security Strategy Study (FS3). The FS3 recommended that, in the absence of near-term options for area defense of the continental United States, other interim options should be adopted, such as partial defenses and anti-tactical missile defenses. Following the Fletcher and Hoffman reports, an interagency panel headed by Franklin C. Miller completed its own review of the new policy.[33] No decision on system architecture had been reached yet, though.

The SDIO

In the meantime, the Strategic Defense Initiative Organization (SDIO) formed in April 1984. Lt. Gen. James A. "Abe" Abrahamson, the first SDIO head, came to be seen by those inside and outside the agency as a charismatic figure with a flair for steering a budget request through Congress.[34] He focused on winning funds from Congress, instead of micromanaging the SDI program. Abrahamson created an organizational culture that featured minimalist managerial teams. The SDIO staff numbered only about eighty by the middle of 1985.[35] The agency quickly earned a reputation as exemplifying the "faster, cheaper, smaller" management approach.

The SDIO operated like a rather large example of an established bureaucratic practice known as a special project office. Previous special project offices had been organized for the *Polaris, Trident,* and *MX* missile programs under the Office of the Undersecretary of Defense for Research and Engineering. These were subject to annual reviews and oversight by the Defense Systems Acquisition Review Council (DSARC). The SDI, however, was free from these constraints. This bureaucratic freedom presented the services with opportunities to keep certain projects. When projects failed to pass DSARC milestones, the services could transfer them to the SDIO budget, while the original service retained project management.[36]

Critics in Congress decried the glut of Washington-based consultants, so-called Beltway Bandits, spawned by SDI, because the SDIO did not have its own laboratories. Rather, the agency contracted its work, spreading SDI funding across the country in laboratories operated by the army, air force, navy, and Department of Energy.[37] Criticism of the Strategic Defense Initiative was not limited to hearings in Congress, but spread out through the media and into public debate. "Star Wars" became a deprecatory and belittling term used by the program's enemies.[38]

Just as Nixon had used Safeguard as a bargaining chip in SALT I talks, Reagan used SDI to leverage arms negotiations. In March 1985, the United States and the Soviet Union started the Nuclear and Space Talks (NST) in Geneva. The United States sought to legitimize SDI and claimed that the Soviet Union had violated the 1972 ABM Treaty. The USSR continued to denounce SDI as an impediment to arms control. At the Reykjavik October 1986 summit talks, the Soviet Union proposed that both sides observe the ABM Treaty for another ten years, and they insisted on retaining the traditional reading of the ABM Treaty, which banned all tests of space-based ballistic missile defense systems and components outside the laboratory. The United States refused these conditions, the main obstacle to an agreement.

In 1987, the Soviets "decoupled" the SDI from treaty negotiations. Ending the program was no longer a prerequisite to an agreement. During September 1987 talks in Geneva, the USSR modified its position further to allow some antiballistic missile research in space. In November 1987, Mikhail Gorbachev told a television audience that the Soviet Union "will not build an SDI, we will not deploy an SDI, and we call on the United States to do likewise." Talks later that year in Washington, D.C., cemented a new relationship between the two countries, and on January 15, 1988, the USSR presented a draft START (Strategic Arms Reduction Treaty) protocol, which committed both countries to abide by the 1972 ABM Treaty for ten years and froze the number of launchers.[39]

Meanwhile, the SDI architecture studies launched in 1984 quietly had concluded in October 1985. The SDIO still called for a development decision in the early 1990s, initial deployment in the late 1990s, and effective service around 2005. One concept actually had no weapons in space. Another had thousands of satellites and seven separate tiers of defenses. The discrepancy signaled how vague the concept of a Strategic Defense Initiative remained. In 1986, a team of Senate staff members on behalf of Sen. William Proxmire (D-WI) interviewed some forty scientists, engineers, and military officials involved in SDI research. They concluded that the SDIO still had no specific systems architectures that could be tested against a realistic set of threat scenarios.[40]

Finally, in September 1987, the Pentagon approved the Strategic Defense System (SDS) Phase I Architecture. It consisted of six major subsystems: a space-based interceptor (SBI), a ground-based interceptor, a ground-based sensor, two space-based sensors, and a battle management system. The SBI, nicknamed Smart Rocks, was a large, garage-like satellite housing ten individual hit-to-kill interceptors. About 300 SBIs would orbit Earth. In case of an attack by Soviet missiles, the SBIs would launch their interceptors at individual Soviet missiles, destroying a large number of them during their boost phase, while they were still pregnant with their multiple warheads and decoys. To a great extent, it was the High Frontier defense system promoted by Daniel Graham.

With the September 1987 approval of Phase I, the Strategic Defense Initiative was on its way to realizing part of the conservative space agenda. Two years later, the first licensed commercial launches would accomplish another key part of that agenda. Both the commercialization and the militarization of space boosted the demand for launchers. The brunt of commercial launches would be on expendable rockets designed and built for the cold war by aerospace firms within the military-industrial complex. The same firms that manufactured those rockets now were selling them on the commercial market. Establishing the Strategic Defense Initiative in space would require a tremendous number of launches. With the shuttle no longer carrying military payloads, could a multi-billion-dollar government project like SDI afford the high cost of riding on expendable rockets? In order to meet its launch needs, the SDIO joined with other Defense Department branches and NASA to develop a new launch system that promised to lower the cost of putting payloads in orbit. The new launcher would be a reusable single-stage-to-orbit vehicle, the National Aero-Space Plane (NASP). NASP, typical of the conservative space agenda, would serve both military and commercial ends. Moreover, typical of the cold war, it would be an expensive, large-scale, long-term program. Although the NASP vehicle was a single-stage-to-orbit concept, the program was the antithesis of the vision that led to the *DC-X*.

Part II / The Quest

CHAPTER 4

X-30: The Cold War SSTO

One of the consequences of the ill-fated flight of the *Challenger* was the removal of military and commercial payloads from the shuttle. The approval of Phase I of the Strategic Defense Initiative (SDI) in September 1987, followed by the first licensed commercial launches in 1989, signified a new and growing demand for launch services. They also represented key points in the realization of the conservative space agenda, which depended on low-cost launch services. As Newt Gingrich pointed out in 1984, "the real gateway to space has been the development of steadily less expensive methods of boosting payloads into orbit."[1] For the commercialization of space, the transfer of expendable rockets built for the cold war to civilian business use settled the question quickly. The same aerospace firms within the military-industrial complex that built the rockets also profited from their commercial use.

The number of annual U.S. commercial launches was far smaller than the number required for the SDI. Creating the SDI would be a multibillion-dollar government project requiring launch services costing additional billions of dollars. Nor was the SDIO alone in need of launch services. Thus, the Defense Department developed a new launch system that promised to lower the cost of putting payloads in orbit. NASA collaborated in the effort because it was interested in a less costly replacement for the shuttle.

The result was another Reagan-era space program known as the National Aero-Space Plane (NASP). NASP would serve the launch needs of NASA and the Defense Department (SDIO included). Thus, NASP also reflected the conservative space agenda. Reagan announced the program in his February 4, 1986, State of the Union Address, as the country was looking for good news about space travel in the wake of the *Challenger* disaster. The "new Orient Express," the president proclaimed, "by the end of the decade . . . [will be able to] take off from Dulles Airport, accelerate up to 25 times the speed of sound attaining low Earth orbit, or fly to Tokyo within two hours."[2] Reagan's speechwriters confused the NASP reusable single-stage-to-orbit vehicle with the Orient Express, a McDonnell Douglas hypersonic aircraft design in which

Federal Express had shown interest. The confusion probably screened the flight vehicle's military mission, though the McDonnell Douglas prototype claimed to be capable of performing either a NASP single-stage-to-orbit or an Orient Express mission, depending on the vehicle's propulsion system.[3]

Although NASP appeared to be different, it was not. It formally began as a classified study, then evolved quickly into a typical cold war effort. NASP was an expensive, large-scale, long-term program. It also was the latest in a series of projects that attempted to make an aircraft do what rockets do, namely, fly into space. The NASP vehicle, known as the *X-30,* would take off and land horizontally like an aircraft, had wings, and even would use a kind of air-breathing engine known as a scramjet. Scramjet is a truncation of supersonic combustion ramjet. Ramjets are jet engines that propel aircraft at supersonic speeds by igniting fuel mixed with air that the engine has compressed. Scramjets are ramjets that achieve hypersonic velocities.

Because of the *X-30*'s conceptual and perceptual resemblance to aircraft, it and similar craft are called aerospace planes or, simply, spaceplanes. Many such vehicles featured a pilot and co-pilot sitting in a cockpit. For obvious reasons, these craft are particularly appealing to the air force. That appeal ultimately paved the way for the undertaking of the NASP program instead of a rival concept known as the *Reusable Aerodynamic Space Vehicle (RASV).* The *RASV* was a single-stage-to-orbit craft that looked and operated like an aircraft, but used rocket power.

The NASP concept was the wrong road. The selection not only stymied development of rocket-powered, single-stage-to-orbit transport, but also propelled the nation into an expensive program that had no chance of success. The Orient Express never happened. The failure of NASP doomed funding for more feasible alternatives. As a result, too, the SDIO did not have an economical launcher for the SDI, and NASA was without a shuttle replacement. Meanwhile, the nascent commercial space industry relied on throwaway technology developed for the cold war. In short, the nation needed to have traveled a different road, one that led to reliable, reusable, reasonably priced access to space.

To Be Like an Airplane

Proposals for spaceplanes were not new. For instance, in a *Popular Science* article published in December 1931, American rocket pioneer Robert Goddard described one (a "stratosphere plane," according to the editors) with elliptically shaped wings and a combination of air-breathing jets and a rocket engine. The rocket engine drove the vehicle when it was outside the atmosphere, while two turbines moved into the rocket's thrust stream to drive two large propellers on either wing, thereby powering the vehicle while in the atmosphere.[4]

Meanwhile, in the Weimar Republic, Eugen Sänger, in his 1933 book on rocket flight, described a rocket-powered suborbital spaceplane, the *Silbervogel (Silver Bird).* Fueled by liquid oxygen and kerosene, the *Silbervogel* theoretically was capable of reaching a maximum altitude of 160 kilometers (100 miles) and a speed of Mach 10. Later, working with mathematician Irene Bredt, whom he married, and a number of research assistants, Sänger designed the *Rocket Spaceplane,* which would launch at a speed of Mach 1.5 from a sled. A rocket engine capable of developing 100 tons of thrust would boost the craft into orbit, where it could deploy payloads weighing up to one ton.[5]

Many in the aerospace field viewed single-stage-to-orbit vehicles as impossible. For some, though, the X-15's excursion into near space made some of these proposed vehicles seem achievable. As a result, single-stage-to-orbit designs occasionally made their way into actual government programs. One example is the *Aerospaceplane,* which bore a generic resemblance to Sänger's *Silbervogel* and Goddard's "stratosphere plane." According to historian Richard P. Hallion, the *Aerospaceplane* was "the first major attempt to develop a true large-scale military logistical spacecraft capable of flying into space and returning to earth in a lifting reentry."[6] The *Aerospaceplane* did not start as a single-stage-to-orbit program, but rather as a highly classified project to investigate a range of launch configurations. Some were single-stage, while others were two-stage designs. Some had rocket engines, while others had air-breathing engines, and still others combined air-breathing and rocket propulsion. The *Aerospaceplane* initially did not call for the development of a full-scale orbital craft, but rather focused on design and construction of a scramjet engine to be launched on top of a missile. Such an engine was being built when the program ended in 1963. Congress had cut funding, and the Pentagon declined to press for its restoration.[7]

The *Aerospaceplane* attempted to imitate an aircraft by flying into orbit. It would take much more than imitation to get a space vehicle to operate like an aircraft, though. The space shuttle illustrates that difficulty. In 1969, George E. Mueller, NASA associate administrator for manned space flight, told an audience that he saw three critical areas for shuttle development. After the engine and the heat shielding came the "checkout and control system which provides autonomous operation by the crew without major support from the ground and which will allow low cost of maintenance and repair. Of the three, the latter may be a greater challenge than the first two."[8]

Mueller's words were prophetic. For the shuttle to have the "checkout and control system" for "autonomous operation," NASA would have to equip it with onboard computers and sensor arrays to monitor and diagnose the shuttle's propulsion and other systems. The space agency had a facility for in-house electronics expertise, the Electronics Research Center in Cambridge, Massachusetts, but closed it in 1970 allegedly as a cost-cutting measure.[9] NASA would have to look elsewhere for its electronics expertise.

In 1968, computerized checkout for aircraft was still in the future. American Airlines was interested in developing computerized checkout technologies for commercial aircraft maintenance. Their maintenance managers told NASA about the difficulty of finding faults quickly and accurately in their complex airplanes. Meanwhile, Pan American World Airways was emerging as an industry leader in this area. In 1970, for example, it was furnishing onboard fault detection and analysis for cockpit instruments and items of flight equipment, including radio altimeter, radio receivers (for navigation and low-visibility landings), transponders (to augment the plane's radar), and electrical generating systems. Pan Am also was extending the use of airborne monitoring systems to detect engine faults. In 1970, the company flight-tested a prototype aboard a Boeing 707. Both American Airlines and Pan Am participated in shuttle design study teams, thereby furnishing each team the advice and experience of the commercial airline industry. American worked with the North American Rockwell team, while Pan Am teamed with McDonnell Douglas and its partners.[10]

While developing the shuttle, NASA had claimed that it would achieve "airplane-like operations," namely, the fast turnaround time typical of commercial airlines. The original turnaround time claimed for the orbiter was ten working days. However, as NASA testified to Congress in 1984, the actual turnaround time on the first flights had been greater than a hundred workdays, but since had been reduced to fifty-three workdays. The agency hoped to reduce that time to twenty-eight workdays by 1988. Although the orbiters' heat shielding tiles required (and still require) nothing less than a "standing army" of thousands, one of the major barriers to cutting orbiter turnaround time in the early 1980s was flight software. NASA had to compile and verify individually the software for each payload, then check the entire payload package in unison. In addition, the ground crew, flight crew, and customers had to verify in real time, and by hand, all parts of the flight software. The House Subcommittee on Space Science and Applications recommended that NASA pay "special attention to substantially reducing the amount of time needed to compile and verify the computer software package required for each mission, or in some way eliminating the need for a new software package for each flight."[11]

The computer and electronic technology needed to endow an orbital vehicle with aircraft-like operations was still in its early stages of development. Engineers talked about computer checkout systems. Health monitoring systems and other current terms had not been invented yet. Onboard computer flight control was a new idea. Not until May 1972 did the first plane, an experimental F-8C fighter, use a computer to control its flight. Computer-controlled flight was more advanced in the design of rocket-powered space vehicles for the military, though. The Boeing *X-20* was to be the air force's piloted Dyna-Soar stub-winged craft boosted atop a *Titan* missile. Had the project continued, the *X-20* would have been the first space vehicle to fly under the control of a computer. For its part, the shuttle was able to take advantage of the experimental work on computer-controlled flight underway at NASA's Dryden Flight Research Center.[12]

Son of Shuttle

A shuttle replacement would be able to take advantage of the dramatic advances in computers and electronics that took place during the 1980s and later. The quest for aircraft-like operations continued within NASA's planning for a shuttle replacement. The focal point of planning studies was at the Langley Research Center in Virginia. There, researchers took single-stage-to-orbit concepts seriously.

Langley's Space Systems Division (now the Space Systems and Concepts Division) under Gene Love supported the space shuttle program by analyzing and testing the aerodynamic design of various single-stage, two-stage, and stage-and-a-half configurations in wind tunnels. Love was particularly interested in single-stage-to-orbit concepts after examining a program called the *Continental/SemiGlobal Transport (C/SGT)*. The *C/GST* would take off, almost attain orbit, then land, delivering people or cargo to any place on Earth in less than two hours. Analysis of the vehicle showed Love that with just a little bit more speed, the *C/GST* could become a single-stage-to-orbit vehicle. With the backing of Stanley R. Sadin, Office of Aeronautics and Space Technology, NASA headquarters, Love split off a little group from the division staff in 1972 to analyze this and other vehicle concepts. Among the division personnel in that group were John Decker, head of the Advanced Technology Section; Beverly "B.Z." Henry, head of the Vehicle Analysis Branch; Paul Holloway, section head; and Chuck Eldred, researcher.[13]

In 1975, Love's small group received some money for two industry studies of single-stage-to-orbit rocket concepts. One team, headed by Rudolph C. Haefeli, came from the Martin Marietta Corporation's Denver Division; the other, run by Andrew K. Hepler, was from Boeing (Seattle). Langley's Chuck Eldred served as technical contact with the Martin Marietta team, while John Decker was the contact for the Boeing team. NASA charged the two teams with determining the future technological development needed to build an operational rocket-powered, single-stage-to-orbit shuttle replacement by the year 1995. Each team analyzed the same three configurations: 1) vertical takeoff; 2) horizontal launch from a sled; and 3) horizontal takeoff with subsequent in-flight refueling. In all three configurations, the vehicle landed horizontally, like the shuttle, and would be capable of flying 500 twelve-hour missions from Cape Kennedy (Canaveral) with a payload of 65,000 pounds (29,500 kg).

Although each team looked at the same vehicle configurations, their basic designs differed. Boeing's was essentially a winged cylinder, while Martin Marietta's had a flattened body. Both teams analyzed vehicles that stored liquid hydrogen in the craft's body and liquid oxygen in the wings, although Martin Marietta also considered vehicles with liquid oxygen in the body, rather than in the wings. Storing liquid oxygen in the wings provided wing-bending load relief for the horizontal takeoff configurations and reduced the overall weight of the vehicle both with and without fuel.[14]

As the study progressed, the Langley, Boeing, and Martin Marietta engineers came to realize that, although reusable vehicles had lower operating costs than throwaway rockets, the real key to lowering operating costs was to reduce the number of people doing operations and maintenance on the vehicle by using what they called (for want of a better name) "operations technology." Researchers would have to develop new hardware and software so that the reusable vehicles could be operated and maintained with far fewer workers. Chuck Eldred explained: "We knew we had to attack the problem of reducing the number of people. We said that making a vehicle fully reusable is a step, but it's not sufficient to get down the cost we're talking about. We've got to do something more."[15]

Both the Boeing and Martin Marietta teams acknowledged how to incorporate aircraft-like operations in a general way. Martin Marietta claimed that they would be able to check out, refuel, and ready their vertically launched single-stage-to-orbit craft in only sixty hours after landing. They would achieve this breakneck turnaround by running operations twenty-four hours a day and by using improved onboard flight systems designed with automated self-test and checkout capability and ground support systems designed with simplified monitoring software.[16] The Boeing team, which claimed that their company had been working on a horizontal-takeoff, single-stage-to-orbit vehicle concept "for over four years," that is, since 1972, aimed at driving down operational costs until the cost of fuel dominated the cost per flight. They also noted that integrating payloads had to achieve the same level of sophistication as commercial air cargo carriers.[17]

The Boeing emphasis on aircraft-like operations originated as much from the Langley shuttle study as it did from the team's leader, Andy Hepler. He had been in a unique position to marry the worlds of rocket and aircraft design. Hepler started at Boeing in the 1950s as an aeronautical engineer working on a variety of aircraft projects, including the B-50, the B-52, and the KB-29 tankers for in-flight fueling. Along the way, he also worked on the *Bomarc* Mach-3 ramjet surface-to-air missile, first launched in August 1958. The *Bomarc* missile had wings and a tail and stood like an airplane ready for vertical launch.[18] He and the *Bomarc* project team were all aeronautical engineers with Boeing aircraft experience. The pool of those knowledgeable about rocketry was limited, while military demand for missiles was expanding dramatically. Between 1956 and 1961, the portion of military orders represented by missile contracts grew from 5.71 percent to 44.35 percent.[19] This demand forced many aeronautical engineers to deal with both aircraft and rockets.

In 1959, Hepler transferred to the Dyna-Soar *(X-20)* program. Launched on an expendable booster, the reusable *X-20* would fly orbital or suborbital trajectories, perform reconnaissance at hypersonic speeds, and land horizontally like an aircraft at many U.S. air bases. Hepler was responsible for stress analysis, structural sizing, and materials selection on Dyna-Soar. He had not yet begun to think about applying airplane operations to rockets; however, he learned about René 41, a high-temperature nickel alloy developed for the *X-20*.[20] Hepler next worked on a mix of space and aircraft projects from the Manned Orbital Laboratory to the SuperSonic Transport (SST), still studying structures.[21] Following cancellation of the SST, Hepler worked for a while on his company's shuttle design before leading a group that examined fully reusable launchers. One of the ground rules that Hepler set for the group, he recalled, was to "get into airplane-type ground ops [operations], because we thought we could significantly reduce general operating costs."[22]

Upon completion of the Langley shuttle replacement study, Hepler and his group had a general design for a cylindrically shaped, delta-winged, reusable, single-stage-to-orbit craft that would take off and land horizontally. Space Shuttle Main Engines would power it, in order to avoid development of a new propulsion system. The craft stored liquid oxygen in its wings and liquid hydrogen in its main body. The airframe and wings would be made of René 41 alloy honeycomb (developed for Dyna-Soar) and an aluminum-titanium honeycomb (developed for the SST). The vehicle used a sled mechanism for takeoff and conventional landing gear for landing.[23]

The Reusable Aerodynamic Space Vehicle

Boeing soon interested the Air Force Space and Missiles System Organization (Los Angeles Air Force Station) in this vehicle concept. The air force dubbed it the *Reusable Aerodynamic Space Vehicle (RASV)* and funded a seven-month preliminary feasibility study in 1976. It concluded (not surprisingly) that the *RASV* was feasible and that it would fulfill air force requirements. The *RASV* would be capable of flying 500 to 1,000 times "with low-cost refurbishment and maintenance as a design goal" from a launch site in Grand Forks, North Dakota, into a polar orbit, or once around the planet in a different orbit, and would be capable of carrying payloads up to 10,000 pounds (4,536 kg) and no larger than 10 feet by 15 feet (3.0 m by 4.6 m). The vehicle would have to reach "standby status within 24 hours from warning. Standby to launch shall be three minutes."[24] The potential customer, the Strategic Air Command (SAC), imposed these requirements.

The *RASV* proposal included a technology development plan as well as aircraft-like operations and "incremental testing of the vehicle through the various flight regimes." The craft also would have the ability to abort when one of its two engines ceased to function.[25] Thus, a single project proposal brought together the ideas of single-stage-to-orbit transport, aircraft-like operations, and incremental flight testing. This was almost the unified project vision that became embodied in the *DC-X* program. The crucial difference is that the *RASV* lacked the "faster, cheaper, smaller" management approach. Instead, it would have been a long-term, large-scale, expensive program typical of those undertaken during the cold war. To go from concept development to testing of the first full-scale *RASV* would have taken ten years and likely would have cost several billions of dollars.[26]

Clearly, the *RASV* had many technological barriers to hurdle. Boeing would have to develop the sled that would accelerate the *RASV* to a speed of 600 feet per second. A Boeing 747 would have to transport the *RASV* back to its launch site (like the shuttle), in case of an abort or a forced landing at a site not equipped with a sled launcher. The actual report section on operations and ground support was exceedingly thin. Indeed, the air force study admitted that ground support was "conceptual in nature. The systems were not subjected to the level of design assessment as was applied to the flight vehicle."[27]

The study offered only a few guidelines on designing for operations and maintenance. Cargo would be "prepackaged and self-sustaining," thereby minimizing cargo loading and unloading operations. The Strategic Air Command insisted that satellites be prepackaged and containerized for quick loading. The design would require minimal checkout at the launch facility with "maximum on-board autonomy and maximum use of an on-board checkout computer system for preflight and postflight operations." In addition, operations and maintenance would decide the selection of subsystems and equipment, "rather than just emphasizing maximum vehicle performance and mission and crew safety only."[28]

In many ways, ground crews would handle the *RASV* like an aircraft. It would be serviced in a B-52 hangar, and an engine could be changed in hours rather than weeks. The *RASV* would use a flight control tower, not a launch control center, and the pilot would make the decision when to launch. Also, the cockpit would look like that of an aircraft, and the pilot would fly it as if it were an airplane. Crew would load payloads from the bottom, not from the top.[29]

However, the only operations or maintenance issue addressed by the feasibility study in any detail was that of rapid refueling. Refueling would have to take place in thirty minutes, which was beyond the limits of available cryogenic technology.[30] With additional funds from the air force, Hepler's group attempted to resolve the problem. They proposed having the *RASV* carry liquid helium to purge and repressurize its fuel tanks at the end of a mission, while the vehicle was still airborne. This solution eliminated the need to purge the tanks on the ground. Rockwell's Rocketdyne subsidiary also contributed to improving vehicle maintenance with a study of a dual nozzle for the Space Shuttle Main Engine, as well as an engine redesign to make it easier to maintain.[31]

Over time, the *RASV* underwent changes in the face of new performance requirements. It now had to be capable of carrying heavier loads into orbit, 30,000 pounds (66,000 kg) once around the globe from any launch site, though most likely from the central continental United States. The craft still had to be able to take off with little notice and have a fast turnaround time, twenty-four or perhaps as few as eight hours.[32] All in all, the air force invested $3 million in the project for technology development. They believed that the *RASV* potentially could provide a manned vehicle that they could place above any point on the planet in less than an hour and that could perform a variety of missions, including reconnaissance and rapid satellite replacement. A reflection of Boeing's confidence in the *RASV* was the December 1982 decision of Chairman T. A. Wilson to give the *RASV* project his approval to propose a $1.4 billion prototype vehicle to the air force.[33]

The *RASV* concept subsequently participated in two military technology studies known as Science Dawn and Have Region. These would be important to the development of future single-stage-to-orbit craft. By proving certain critical technologies, they convinced key people of the feasibility of a single-stage-to-orbit vehicle. As a rule, technological breakthroughs stimulated fresh belief in single-stage-to-orbit transport. Science Dawn, the precursor to Have Region, took place between 1983 and 1985 and involved Boeing, Lockheed, and McDonnell Douglas. Science Dawn's objective was to analyze manned single-stage-to-orbit vehicles that took off and landed horizontally and that could deliver up to 10,000 pounds to a polar orbit for the Strategic Air Command. Industry would construct the vehicles from off-the-shelf technology. The craft would be capable of flying two missions per day, have a turn-around time of twelve hours, be ready to fly within two hours of an alert, and be able to remain in orbit for twenty-four hours. Essentially, this was the *RASV* concept.

Lockheed named its Science Dawn vehicle the *Zero Length Launch*

TransAtmospheric Vehicle (ZEL-TAV). McDonnell Douglas called its vehicle the *Global Range Mach 29 Aerospace Plane (GRM-29A)*, while Boeing proposed its *RASV*. The Science Dawn study concluded that single-stage-to-orbit vehicles were possible using existing metallic alloy technologies, but they represented a high risk. The Science Dawn final report recommended initiating a program to build and test metallic structures for single-stage-to-orbit vehicles. The testing of the structures took place from 1986 through 1989 as the classified Have Region program.[34]

In Have Region, the same three industry participants (Lockheed, McDonnell Douglas, and Boeing) designed and built structural cross sections and evaluated their capability of being manufactured, their structural strength and weights, and their thermal performance. Have Region sought to develop the capability to build an operational vehicle within four years, that is, in 1990. The effort, with a budget of about $35 million, included testing titanium structural cross sections under simulated ascent and reentry conditions with a thermal protection system, and testing pressurized fuel tanks under the same conditions of temperature, pressure, and thermal loads. Basically, too, it was a test of Boeing *RASV* structural materials.

One structure studied was a full-scale cross section of a single-stage-to-orbit airframe. The other two cross sections were about 40 percent of full scale. All structures incorporated integral cryogenic tanks that would contain the liquid hydrogen or oxygen (like the *RASV* and the Langley study vehicles). Not all structures passed the tests, but the failures arose from such factors as bad welding and test equipment failures. In other words, the results were favorable, even the failures. Have Region concluded that one could build metallic single-stage-to-orbit spaceships, but that the use of modern composites had the potential to deliver much more robust and operable single-stage-to-orbit vehicles.[35] Nonetheless, Have Region did not test composite material structures.

Science Dawn and Have Region gave momentum to the development of a rocket-powered, reusable, single-stage-to-orbit craft. That momentum dissipated, though, when it was overcome by a stronger force, one that sought to build a reusable orbital craft that flew with an aircraft engine. The conceptual resemblance of such a vehicle (the spaceplane) to an airplane (complete with air-breathing aircraft engine) appealed to the air force and its pilot mentality. Jet engines and wings are technologies that air force pilots and officers understand. Not surprisingly, then, as the air force began studying shuttle replacement concepts in its *TransAtmospheric Vehicle* program, air-breathing jet engines came under consideration.

To Breathe Like an Airplane

The air force's *TransAtmospheric Vehicle (TAV)* program dated back to January 1982, when the Aeronautical Systems Division of Wright-Patterson AFB initiated an internal study of shuttle successor designs. Planning soon expanded, and the designs took on the generic name *TransAtmospheric Vehicles,* because the air force hoped that they could operate equally well in the atmosphere and in space. The *TAVs* could take off from an existing aircraft, such as a Boeing 747, or use an entirely new launch vehicle developed specifically for it, or be single-stage-to-orbit craft. The Aeronautical Systems Division considered each architecture in depth.[36]

Phase I *TAV* concept studies began in May 1983 with an award to Battelle Laboratories, Columbus, Ohio, which in turn worked with Boeing, General Dynamics, Lockheed, and Rockwell to derive appropriate design concepts. McDonnell Douglas did not participate, but rather used its own money to develop a *TAV* concept. The result was fourteen different *TAV* designs and no agreement on the type of engine the *TAV* should use: a conventional engine, such as a derivative of the Space Shuttle Main Engine; or a complex propulsion system that combined air-breathing and rocket engines.[37] Those responsible for selecting among the *TAV* designs essentially could not choose between a rocket and a spaceplane.

By the time Phase II began in August 1984, the *TAV* had grown into a major air force study and had the potential to become a significant agencywide or even an interagency project. The Wright-Patterson Aeronautical Systems Division drew on other air force branches, such as the Strategic Air Command and the Space Command, to define *TAV* mission requirements and operational concepts. NASA and navy representatives sat on an upper-level steering group. This expanded interest in *TAV* resulted directly from the SDI and space station Reagan initiatives, both of which boosted launcher needs.[38]

After the Aeronautical Systems Division established a *TAV* project office under the direction of Lt. Col. Vince Rausch in December 1984,[39] several external developments intervened to influence the subsequent course of *TAV* work and eventually led to its transformation into the NASP program in January 1986. As *TAV* rocket-powered, single-stage-to-orbit concepts gained followers and program dollars, interest grew among enthusiasts of scramjet hypersonic flight, the same engine type that would be proposed for the National Aero-Space Plane.

The air force, NASA, and the navy all had initiated separate studies of scramjets as a hypersonic propulsion system. Generally, though, the air force and NASA envisioned scramjets propelling manned vehicles, while the navy's interest lay primarily in missile propulsion.[40] Meanwhile, the Advanced Research Projects Agency (ARPA) had initiated Copper Canyon, its own classified scramjet study, in early 1984. Colonel Rausch simultaneously served as director of both Copper Canyon and the *TAV* project. In creating Copper Canyon, ARPA had in mind the Aerospaceplane program of the 1960s, specifically a single-stage-to-orbit craft powered by air-breathing engines. In June 1985, ARPA extended the Copper Canyon study and issued airframe study contracts to Boeing, Lockheed, and McDonnell Douglas and engine contracts to Marquardt and Aerojet General. Copper Canyon now had three phases. The work already performed became Phase 1. A period of technology maturation would make up Phase 2, while Phase 3 would consist of fabrication and test flights of a vehicle.[41]

Following President Reagan's State of the Union Address calling for the creation of the Orient Express, and in compliance with a National Space Security Directive, NASA and the Department of Defense undertook the joint National Space Transportation and Support Study, completed in May 1986. The study endorsed a variety of launcher types, including single-stage-to-orbit and air-breathing vehicles. It also recommended giving the highest national priority to the technological advances required for spaceplane development and flight testing.[42]

The Air Force *TAV* and the ARPA Copper Canyon projects now merged into a new, larger program comprising the myriad government agencies involved in hypersonic, air-breathing engine studies at one time or another. The title National Aero-Space Plane (NASP) replaced all earlier designations on December 1, 1985. The following month, on January 1, 1986, the Air Force Systems Command announced establishment of the NASP Joint Program Office (JPO) at the Aeronautical Systems Division, Wright-Patterson AFB. The NASP JPO included air force, navy, and NASA personnel. NASP's Phase 1 was the completed Copper Canyon work. Phase 2 began with the creation of the Joint Program Office.[43] With the start of the NASP program, spaceplane enthusiasts triumphed over the rocket-powered *RASV*.

The National Aero-Space Plane

The National Aero-Space Plane program proposed to design and build two *X-30* research aircraft, at least one of which would achieve orbit by flying in a single stage through the atmosphere at speeds approaching Mach 25. The *X-30* would use a multicycle propulsion system. It would begin with a low-speed subsonic jet, switch to a midspeed ramjet, and finally convert to a scramjet, with the progression from one to the next occurring during flight. For fuel, the NASP *X-30*s would use a nearly frozen form of liquid hydrogen. The so-called slush hydrogen also would cool the craft, which would burn the hydrogen with oxygen scooped from the atmosphere.[44]

While the NASP program gathered together several existing government research projects, the original vehicle configuration came from Tony Du Pont, although his vehicle was not a lifting body like the NASP *X-30*. Du Pont had his own company, Du Pont Aerospace, but also had been the Air Research Manufacturing Company program manager on the hypersonic research engine (HRE) project. On several of his frequent visits to NASA's Langley Research Center, Du Pont tried to sell a proposal for an aircraft that would take off from a runway and fly up to Mach 15 with large scramjets mounted underneath. Langley analysts told Du Pont that his podded scramjets always would have more external drag than the thrust produced by the engines. Therefore, rather than mounting the engines on the outside, like on an aircraft, Du Pont would have to integrate them into the airframe to reduce external drag. Langley researchers also found his proposed structural weight to be unrealistically light. Undaunted, Tony Du Pont designed a new flight-test vehicle, a single-stage-to-orbit craft powered by a scramjet integrated into the airframe, and submitted it as an unsolicited proposal for funding to the Advanced Research Project Agency (ARPA). Bob Williams, an ARPA manager, liked Du Pont's proposal enough to fund a more detailed analysis, and dubbed the new program Copper Canyon.[45]

ARPA gave Du Pont a contract for computer studies of engine cycles. Not surprisingly, his computer models showed that his multicycle scramjet engine concept would be able to power a single-stage-to-orbit craft. The computer models, however, rested on a number of highly questionable assumptions, optimistic interpretations of results, and convenient omissions. For example, his 1983 design had no landing gear, nor did it have any room for error or growth in vehicle size or weight, which are normal givens in launcher development. His design also featured a wing structure that ran through the center, load-bearing liquid hydrogen tank, a questionable arrangement.

During Phase 2 of NASP, government and contractor researchers independently discovered that the integrated wing and tank structure could not sustain the temperature variations transmitted by the hot wing. Even earlier, between 1984 and 1986, as part of Copper Canyon, Marquardt and the General Applied Science Laboratory attempted to replicate Du Pont's engine study, but failed to reproduce his results. Government and industry researchers working on Phase 2 of NASP also failed to replicate Du Pont's design results using methods commonly accepted by aerospace practitioners.[46] In 1986 and 1987, the eight companies granted airframe and engine contracts, independently and in competition with one another, concluded that the baseline design would not fly. Each contractor quietly began to design its own NASP vehicle.[47]

Ivan Bekey, former director of advanced programs at NASA Headquarters' Office of Space Flight, characterized NASP as "the biggest swindle ever to be foisted on the country." Bekey was an enthusiast for, not a critic of, single-stage-to-orbit concepts. The NASP program was "highly classified," according to Bekey, "and it was full of dubious aerodynamic claims and engine-performance claims and structural and thermal claims that were prevented from being scrutinized by the classification." He judged that the project's assumptions were "very extremely optimistic." NASP "made no sense." "I started penetrating the NASP program," Bekey explained, "and I found out it was full of hot air. It was really a fork in the road, here, in your thinking. If one wanted to take on a program as risky as the NASP, then one shouldn't rule out an all-rocket, single-stage-to-orbit program, which is far less risky. If, on the other hand, one believed that the technology for NASP was not so risky, then we could build a prototype right away."[48]

NASP, nonetheless, had its supporters. Even if Du Pont's claims were overstated and too optimistic, some hoped that the program would help to develop critical technologies and to advance scramjet research. The military advantages and uses of a hypersonic aircraft would be tremendous. More exciting, perhaps, was the prospect of lowering the cost of getting into space. An air-breathing, single-stage-to-orbit spacecraft with reusability and the turnaround of an aircraft could be operated for far less than other launchers, NASP supporters argued. Du Pont, in fact, had claimed airplane-like operations for his launcher design. NASP was expected to resemble a standard aircraft in its ground and control personnel demands, perhaps with a total crew numbering in the dozens, not the thousands required for the space shuttle, which also was supposed to have lowered launch costs. A survey of NASP's most important technical achievements, however, omits any mention of operations or maintenance.[49]

NASP also failed to utilize small management teams. In the words of program manager Vince Rausch: "There was a hope that NASP would be cheaper, better, faster."[50] Many within the program thought that the Pentagon had assured NASP a "fast track" that minimized red tape. The program was neither inexpensive, fast, nor lean, however. The NASP Joint Program Office (JPO) started rather small, with fewer than a hundred people in 1987 (up from twenty-five the previous year). The JPO soon fattened into a multi-layered office that resembled a typical air force procurement program, and in 1990 it had plans to swell still more. Organizing the contractors into a team complicated the management structure further, and matrix management, which the air force commonly used, soon followed.[51] The project was fated to become even larger and more complex.

In 1986, Congress cut fiscal 1987 NASP funding as part of an overall effort to cut spending in the face of an enormous deficit whose reduction had been mandated by the Gramm-Rudman-Hollings Act (the Balanced Budget and Emergency Deficit Control Act). The smaller NASP budget meant fewer resources for all eight contractors, though Rocketdyne participated at its own expense. On October 7, 1987, the JPO eliminated Boeing, Lockheed, and General Electric from the project based on the results of program reviews. McDonnell Douglas, General Dynamics, and Rockwell would proceed with advanced airframe studies.[52]

By 1988, the three remaining contractors began to arrive at their basic NASP vehicle designs. The JPO concluded that orbit could not be achieved without the use of a rocket engine. Program personnel also grasped that the design competition was consuming available resources at an astounding rate. The program could not afford the luxury of funding multiple vehicle designs. It was time to pick a single design. Two different approaches emerged. One, supported strongly by NASA and the air force, favored a standard competition leading to the selection of a single airframe manufacturer and one propulsion company. The other, favored by some contractors, supported forming all participants into either a joint venture or a cooperative arrangement. Proponents argued that the government would get the best results from each contractor, and the program would gain cost savings by eliminating contractor redundancies.[53] In the end, and only gradually, the JPO formed a contractor team and settled on a single design that included the best of all the contractors' ideas.

As discussions continued between the JPO and contractors, two new approaches to team formation arose. In one, a single contractor would oversee the efforts of the other participants, receive a larger share of the contract and work, and dominate the design. The other approach would retain the team idea, but a competitive Phase 3 source selection would determine which would become the prime contractor. The JPO decided to form a single contractor team, which would collectively select a vehicle configuration. Selection of the team leader was left to the contractors, who favored the recommendation of the head of the JPO.[54]

All in all, NASP was a failure. Despite the predictable difficulties with the multicycle scramjet engine, some NASP research led to important advances in slush hydrogen, composite materials (carbon-carbon in particular), titanium alloys (especially Beta 21S), and certain avionics subsystems.[55] There was nothing "faster, cheaper, smaller" about the NASP program. Its elaborate and cumbersome managerial structure reflected the multitude of government agencies within the military-industrial complex with an interest in developing hypersonic scramjet technology and a single-stage-to-orbit craft: NASA, ARPA, the air force, the navy, and SDIO. Moreover, an impressive range of aerospace industry contractors contributed to the effort: Boeing, Lockheed, McDonnell Douglas, General Dynamics, Rockwell, General Electric, Pratt & Whitney, and Rocketdyne. In short, NASP was a typical cold war big technology program.

NASP also was the nation's single-stage-to-orbit program. As such, it represented a certain institutional belief and confidence in the feasibility of single-stage-to-orbit transport. NASP also managed to develop new materials possibly useful for future space projects. Its failure and eventual termination demonstrated unmistakably that an air-breathing, single-stage-to-orbit craft was not the road to travel. For many, NASP's failure also cast doubts on the feasibility of any single-stage-to-orbit project, and it hampered efforts to fund feasible rocket-powered alternatives. It also left the SDIO and NASA without an economical launcher. The country needed a program to research, develop, and test the technologies needed for single-stage-to-orbit transport. Instead, it got a big expensive failure.

The NASP program also showed that those in charge of developing the nation's launchers had failed to learn the lesson of *Challenger.* The shuttle had been the nation's single carrier of satellites into orbit. That calamity showed the folly of relying on a single system to launch all defense, NASA, and commercial payloads. The attempt to create another single vehicle for NASA, the SDIO, and the air force repeated the mistake of past policy. Not only that, but NASP also was supposed to have contributed the Orient Express for business travelers to Japan. Policymakers obviously had *not* learned the lesson of *Challenger.* The financial, materiel, and human resources expended on NASP could have been, and should have been, employed in a wiser pursuit of several competing space and hypersonic vehicle concepts.

The failure of NASP also was a failure to develop the operational and maintenance technologies needed for a true reusable single-stage-to-orbit spaceship. The program talked about the importance of aircraft-like operations, but undertook no significant initiatives to develop them. Their development, instead, had been left to industry, as had become the practice with NASA and the Defense Department. Without a contract or other incentive, why would industry develop aircraft-like operations on its own and at its own expense? Finally, despite all the talk about aircraft-like operations, no agency proposed to demonstrate aircraft-like operations with an actual vehicle. Advocates promoted and praised aircraft-like operations as a cure for lowering launch costs, but achieving them seemed to lie in the world of mythology, like Don Quixote's Golden Helmet of Mambrino.

CHAPTER 5

Space Visionaries

The idea of a reusable, rocket-powered, single-stage-to-orbit vehicle is far from new. The imagination of science fiction authors Jules Verne, Hugo Gernsback, and their intellectual heirs have provided a vision of single-stage-to-orbit transport for over a hundred years. Jules Verne portrayed such a vehicle in *From the Earth to the Moon direct in ninety-seven hours and twenty minutes, and a trip around it* (*De la terre à la lune; trajet direct en 97 heures*), first published in 1865. It also appeared in English as *The Baltimore Gun Club*. Verne's trip to the Moon took place in a bullet-shaped, single-stage-to-orbit craft shot from an enormous cannon near Tampa, Florida (while sites in Texas competed for the honor).[1] As is usually the case, human imagination outran technological development.

Aerospace engineers have produced a number of single-stage-to-orbit concepts. These have arisen out of a combination of their imaginations and the technological breakthroughs that inspired them. Like the concepts of science fiction writers, these did not become government programs. The aerospace community within the military-industrial complex rejected them as not feasible. Like the concepts of science fiction writers and imaginative aerospace engineers, the original idea for what became the *DC-X* and its SDIO program started neither as a government program nor as an agency requirement, but in the mind of an individual. Unlike other rocket single-stage-to-orbit ideas, though, this one became a government program. In order to understand how and why that happened, we first must examine some of the many failed ideas for single-stage-to-orbit rockets and the role played by technological breakthroughs.

Thinking Inside the Box

Typical of the earlier single-stage-to-orbit rockets were two proposals that NASA received as space shuttle proposals. One was an ambitious design from the Chrysler Space Division for an alternate space shuttle concept submitted in 1970. Chrysler's *Single-stage Earth-orbital Reusable Vehicle (SERV)* resembled the Apollo command capsule. Its payload bay measured 90 feet (27.5 m) across by 66.5 feet (20 m) high and could carry 116,439 pounds (about 53,000 kg) to low-earth orbit. Chrysler proposed a 12-module aerospike engine fueled by liquid oxygen and liquid hydrogen and capable of developing an impressive 5.4 million pounds (24 million newtons) of thrust. The aerospike was the kind of engine NASA's Marshall Space Flight Center did not want to use for the shuttle, because its development time would be too long.[2] *SERV* received little publicity, and even less official support, probably because it was too technologically daring at a time when budget realities were pushing NASA farther and farther away from reusability, as well as away from two-stage-to-orbit designs.

The other single-stage-to-orbit shuttle concept came from Edward W. Gomersall of NASA's Office of Advanced Research and Technology at Moffett Field, California (now the Ames Research Center) in 1970. Gomersall drew upon analysis conducted internally and under contract by General Electric and Boeing. His vehicle used "near current state-of-the-art structures, thermal protection systems, and propulsion." Nonetheless, Gomersall saw the need for more work on his single-stage-to-orbit ship. "Major problem areas are the flow and thermal protection in the engine area and terminal maneuvering and landing." The vehicle, which stood 152 feet (46 m) tall and was 63 feet (19 m) in diameter at its widest point, took off and landed vertically. Powered by a set of aerospike engines under development by Rocketdyne,[3] the craft essentially borrowed from the *Saturn V* and could undertake the same payload weight missions as that rocket or the NASA shuttle as then conceived.[4]

While the unproven and never built aerospike engine would propel the Chrysler and Gomersall vehicles, the powerful Space Shuttle Main Engine encouraged other aerospace engineers to propose single-stage-to-orbit spaceships. In 1973, two engineers, one of whom, Rudi Beichel, had worked at the Nazi Peenemünde plant, reported that "several key developments which have matured within the last few years may have brought us within striking distance of the natural solution to this problem: fully reusable one-stage-to-orbit shuttles." The authors were referring to the Space Shuttle Main Engine. They readily admitted that "the reusable one-stage-to-orbit mission has long been recognized as a technical challenge of high order."[5]

A recurring theme in the history of single-stage-to-orbit launcher evolution is the argument that technological breakthroughs have made single-stage-to-orbit transport feasible. Technological breakthroughs were the driving force behind the conversion of many "true believers" to the idea of the practicality of single-stage-to-orbit craft. Although NASA disregarded single-stage-to-orbit concepts for its shuttle, and the air force shut down the Aerospaceplane program, a few individuals in the aerospace sector of the military-industrial complex persistently imagined and proselytized single-stage-to-orbit possibilities. Technological advances inspired their visions, and in turn their visions excited the imaginations of others, but not those in control of government or business programs. One such technological dreamer was the late Phil Bono.

For Bono, the technological breakthrough was the third stage of the *Saturn V* rocket. An aeronautical engineer specializing in structures and propulsion just out of college, Phil Bono went to work for North American Aviation in 1947, analyzing Nazi *V-2* missiles. He later spent nine years at Boeing on the Dyna-Soar program, then worked for the Douglas Space and Missiles Company as technical director on NASA's Apollo and Voyager projects. As Douglas director of unconventional launch vehicle studies, then director of large launch systems, Bono became involved in studies of reusable vehicles.[6]

Bono realized that the third stage of the *Saturn V* rocket, the Douglas-built *S-IVB,* was light enough to be used as a single-stage-to-orbit vehicle capable of putting an 8,000-pound (3,600-kg) two-man *Gemini* space capsule into orbit. He redesigned the *S-IVB* as part of two internal Douglas studies completed in November 1966. He called it the *Saturn Application Single-Stage-To-Orbit (SASSTO).* It was 62 feet (19 m) tall, including the optional *Gemini* capsule. It landed vertically on four rigid legs and burned liquid hydrogen and oxygen. The major objection to *SASSTO,* Bono felt, was not technological but psychological—the need for the craft to have wings. Bono conceived a variation on the *SASSTO* called *Saturn Application Retrieval and Rescue Apparatus (SARRA).* It also was a reusable, single-stage-to-orbit craft. *SARRA*'s main functions were to supply space stations in low orbit with payloads of up to 8,100 pounds (3,670 kg) and to act as a "space ambulance" to bring sick or injured astronauts back to Earth.[7]

Bono also designed the *ROMBUS* single-stage-to-orbit vehicle for NASA. In 1962, the space agency contracted with Douglas to conduct the two-year Post-Saturn Launch Vehicle Study to identify the next generation launcher after the expendable *Saturn.* Bono headed the study. The resultant *Recoverable Single-Stage Spacecraft Booster* was the subject of two patents, No. D201,773 in

July 1965 for the configuration and No. 3,295,790 in January 1967 for the detail design, awarded to NASA on behalf of Phil Bono. Bono redesigned it as a freight-carrying craft bearing the name *ROMBUS (Reusable Orbital Module-Booster and Utility Shuttle).* It was a heavy, huge launcher. It stood 95 feet (29 m) tall with a base diameter of 80 feet (24 m) and weighed 14 million pounds (6.35 million kg), including a payload of 400 to 500 tons (362,000 to 453,000 kg). Bono believed that *ROMBUS* would be highly maneuverable, capable of being reused "perhaps a hundred times," and have an "average turn-around time . . . [of] only one and a half weeks."[8] To many, the *ROMBUS* looked like something out of science fiction. Indeed, according to the late science fiction writer and aerospace engineer G. Harry Stine, the *ROMBUS* was the inspiration for Disneyland's former "Flight to Mars" ride.[9]

Bono and other aerospace engineers who proposed single-stage-to-orbit concepts drew their inspiration from technological changes created by and for the military-industrial complex (the Space Shuttle Main Engine, the aerospike engine, the third stage of the *Saturn V* rocket). The list of those developing single-stage-to-orbit designs would not be complete without a mention of the work of Gary Hudson. His inspiration did not come from technological breakthroughs as much as from his personal and ideological (libertarian) conviction of the need to develop economical launchers for the commercial market, not the military-industrial complex. This belief drove Hudson even into the 1990s, when the firm he founded, the Rotary Rocket Company, was developing a manned single-stage-to-orbit rocket, the *Roton ATV (Atmospheric Test Vehicle),* entirely with private funding from novelist Tom Clancy and other investors.

The Phoenix

Hudson, an award-winning high school science student in St. Paul, Minnesota, studied physics, astronomy, and microbiology at the University of Minnesota. He left the university in 1971, without receiving a degree, and began a career as a writer, consultant, and lecturer on private sector approaches to space. He started a nonprofit organization called the Foundation Institute, inspired by Isaac Asimov's science fiction trilogy of the same name, to promote the development of space technology. He also published the *Foundation Report* (later issued as the *Commercial Space Report*), featuring articles about future uses of space.[10]

Following his unsuccessful venture with the *Percheron,* Hudson abandoned expendable launchers to focus on reusable vehicles. In 1982, he founded a new company, Pacific American Launch Systems, Inc., to design and build a family of small single-stage-to-orbit vehicles known as the *Phoenix,* which Hudson had begun working on during the 1970s.[11] Max Hunter was a senior vice president in the firm and was to have "complete technical authority" for the development of Pacific American's single-stage-to-orbit vehicle.[12] Hudson believed that the *Challenger* disaster would benefit single-stage-to-orbit concepts. "The myth of NASA space superiority was shattered by the loss of the *Challenger* and subsequent missteps," he wrote. As a result, "more people were willing to take the concept of Phoenix seriously."

Early *Phoenix* designs dating from the 1970s assumed that lightweight airframes and engines (most of the structures were aluminum or steel) were more important than propulsion performance. As with his other launcher projects, Hudson intended *Phoenix* to be developed, tested, and operated commercially with private funds. The Pacific American *Phoenix,* whose development began in 1982, was actually a family of small and large launchers that took off and landed vertically. As the designs of the larger launchers evolved, Hudson changed the choice of fuel, settling on liquid hydrogen and oxygen; the thermal protection system changed from passive to active (using water to cool the bare aluminum skin); and aerospikes replaced the traditional bell nozzle engines. The smaller *Phoenixes* were rather different. They used bell nozzle engines and aeroshells (and possibly fuel tanks) made of composite materials.[13]

Hudson initially envisioned *Phoenix* customers as coming from both the commercial and military markets.[14] He reportedly began offering his vehicle designs to the Air Force Space Division, Office of Advanced Plans, as early as 1983.[15] The *Phoenix* launchers were especially applicable to air force needs, Hudson argued, because they could fly payloads of almost 20,000 pounds (about 9,100 kg) into low orbit without a crew, or deliver 1,000 pounds (about 4,500 kg) with a crew. An automated vehicle checkout system and simplified design meant the military *Phoenix* needed a crew of only five for launch, and the vehicle could be ready for takeoff in less than two hours, in any kind of weather.[16]

In 1985, Hudson announced that Pacific American had signed an agreement with Society Expeditions, Inc., of Seattle to provide charter space-tourism flights to low orbit aboard the *Phoenix-E* starting no later than 1991.[17] In

1989, Hudson tried to sell the Space Development Corporation, a private firm he had created, to Lt. Gen. James A. Abrahamson, then director of the SDIO. Max Hunter would have been the firm's chief engineer. As its first project, the Space Development Corporation would have funded Pacific American to develop the *Phoenix* single-stage-to-orbit spaceship.[18] Hudson again tried to sell his *Phoenix* single-stage-to-orbit concept to the Air Force Space Division in early 1986. Although he came close to obtaining a study contract, he was unsuccessful, and it was his last attempt to ferret out government funds for the project.[19] Following that venture, he returned to the private sector as the source of funding.

Rejecting Queen Isabella's Advisors

It is an exceptional and creative individual whose mind can extend beyond the paradigm or "box" accepted by his or her peers. One such engineer was Max Hunter. His innovative thinking about single-stage-to-orbit transport is central to our story. Hunter did not begin his career in the space industry, but in the aircraft business. This one fact had a significant bearing on the development of his thinking about space transportation. After graduating from the Massachusetts Institute of Technology in 1944 with an M.S. in aeronautical engineering, he joined the Douglas Aircraft Company at its Santa Monica, California, plant. Hunter worked as an aerodynamicist on a variety of commercial aircraft, such as the DC-7 and DC-8, the Santa Monica plant's specialty, and bombers, including the XB-42 and XB-43. At the time, operating costs came under the purview of the aerodynamic performance group. Consequently, Hunter kept track of the operating costs of all Douglas Aircraft and competitors' commercial airplanes, using the Air Transport Association's operating-cost-calculating formula.[20]

The Air Transport Association maintained a mathematical model that simulated the operational costs of different types of aircraft flown commercially. The model originated with two United Air Lines engineers who published an article in 1940 that set up a mathematical model for DC-3 operating costs and presented the parameters that gave a good estimate of the cost experienced in actual operations. Their model proved so useful that the Air Transport Association took it over and maintained it by making use of the actual operating cost data of all the airlines.[21]

In 1951, when Douglas entered the missile field, Hunter became head of missile aerodynamics and oversaw the design of the *Nike-Ajax, Hercules,* and other missiles. As chief missiles design engineer, he oversaw the design of the *Thor, Nike-Zeus,* and other missiles, then as chief engineer of space systems, he was responsible for such Douglas space efforts as the *Thor* (known commercially as the *Delta*) and the *Saturn S-IV* stage.[22] Following the launch of *Sputnik,* Douglas formed an internal space committee with representatives from each of the firm's plants to brainstorm ideas about space transport, whether propelled by solid- or liquid-fueled rockets or by nuclear engines. Hunter served as head of the committee. The enormous potential power of a nuclear engine convinced Hunter and the space committee that a reusable, single-stage-to-orbit vehicle was possible, if it were powered by a nuclear engine. In fact, they thought that the vehicle might even be capable of flying from Earth to the Moon, much like the spaceship proposed by Jules Verne's fictional *Baltimore Gun Club.*

Douglas dubbed the nuclear-powered rocket the *Reusable Interplanetary Transport Approach (RITA).* According to a 1961 Douglas Missiles and Space Systems Engineering report, *RITA* promised "economy of operation, convenience in terms of flight schedule, ease of maintenance, and versatility of utilization." "The real reason for building nuclear space transport vehicles," the report argued, was "their ability to achieve low cost space transportation."[23] Here again was a new engine technology sparking single-stage-to-orbit concepts. This was thinking inside the box.

Hunter, like many aerospace engineers and managers, realized that spaceflight was going to be enormously expensive. He also wondered if spaceflight would always have to be expensive, and how low the cost might be brought down. He also noticed that *RITA* somewhat resembled an airplane. A novel, but simple, idea occurred to him: Why not apply the Air Transport Association's operating-cost equations to the rocket? Most of those involved in rocketry had never heard of the Air Transport Association and had no idea how the world of airplanes calculated costs.[24] "I am presumably," Hunter wrote in 1972, "not the only person who made most of his life in rocketry after first having a strong involvement with the economics of air transportation, but subsequent events have convinced me that it is a pretty rare combination."[25]

Consequently, an essential aspect of the *RITA* spaceship was its marriage of commercial aircraft economy of operations, fast turnaround, and ease of maintenance to a rocketship. In order to lower launch costs, Hunter maintained, a vehicle had to be not only reusable, but it had to operate like an aircraft. This was the same conclusion reached by Andy Hepler and the members of the Langley shuttle replacement study. "We were completely unsuccessful in selling either nuclear rockets or low-cost space transportation then, although a real try was made," Hunter wrote. "At the time it seemed to be that we were never able adequately to allay people's fears about the ground-based launching of nuclear rockets." Later, Hunter came to realize that other factors were involved. "The economic ideas were too foreign for the rocket (weapon) engineers and scientists of the day."[26]

Reducing launch costs was not a new idea. Aerospace engineers and managers knew that reusable launchers would save money. If a rocket can be used two or more times, it will cost less per launch than a rocket that can be used only once. Imagine, for example, having to purchase a new automobile for every new trip versus purchasing an automobile once and driving it thousands of times. It is obvious that the cost per trip is less. Realizing that space transportation costs can be calculated like airplane costs is rather different and requires an understanding of both aircraft and rockets, such as that acquired by Max Hunter at Douglas. That realization led Hunter to propose applying aircraft operations, fast turnaround, and ease of maintenance to the nuclear-powered *RITA*.

A few other aerospace engineers and managers also began to realize that one could compare aircraft and rocket costs. For example, in 1967, researchers at NASA's Flight Research Center (now NASA Dryden) offered the X-15 operational experience as a basis for analyzing the cost of flying a reusable spacecraft. Like a reusable spacecraft, each X-15 flew about sixty times over a period of more than seven years at near-space altitudes and returned to a conventional landing strip.[27]

Speaking at a 1969 symposium on reducing launch costs, Frank J. Dore, Convair Division of General Dynamics, considered development and construction, not operational, costs. He pointed out the illogic in launchers' being more expensive than aircraft when most aircraft were far more complex.The main difference between launchers and aircraft, he argued, was that the latter were designed to be flown over and over with minimal maintenance between flights. The two reasons why aircraft were cheaper to develop and operate than launchers, he noted, were that "the aircraft is designed to be developed in an incremental manner, i.e., taxi tests, then low-speed flight, and then faster and faster flights to explore the full flight envelope in a safe, controlled manner.

Two, the malfunctions that occur during the development and operational life of the aircraft are seldom catastrophic since the systems are designed with sufficient redundancy (at the expense of increased complexity, weight, and cost) that the pilot can either accomplish the mission or recover the expensive aircraft and payload."[28]

These comments bespeak an increasingly sophisticated understanding of the difference between reusable and expendable rockets among some of those involved in designing space transport. Nonetheless, the majority of those designing rockets did not see such possibilities. "Almost everyone convinced themselves that space would forever be an expensive thing," Hunter reflected. "Each did his own back-of-the-envelope calculations to prove that rockets were large, required heavy fuel loads, and would always be expensive. Just like Queen Isabella's advisors, all the backs of the envelopes were wrong. But only a handful of us knew that in 1960, and we were not able to make the points hang in there."[29]

Star Clipper

Hunter's thinking continued to evolve. In January 1962, shortly after completing the *RITA* report, he moved to Washington to serve on the National Aeronautics and Space Council, an interagency advisory group on national space policy, under Presidents Kennedy and Johnson. In 1965, he left the council to take a position with the Lockheed Missiles and Space Company, in Sunnyvale, California, designing missile systems. Among numerous other projects, he conceived, started, and ran Lockheed's shuttle program, the stage-and-a-half *Star Clipper*.[30]

Hunter continued to pursue two interrelated ideas: reducing launch costs and making rockets more like aircraft. While still on the National Aeronautics and Space Council, he wrote what he described as "a rather wild paper in early 1964 (once again) to delineate the startling possibilities of nuclear propulsion."[31] He continued to reflect on the problem and came up with the stage-and-a-half concept, eventually known as the *Star Clipper*. "The first day I went to work at Lockheed in the fall of 1965," Hunter recalled, "I was asked if there was anything on my mind which I thought should be done that wasn't being done. Ten days later I produced a Lockheed internal memorandum on orbital transportation." The date of the interdepartmental memorandum was October 28, 1965.[32]

"For the sake of general rabble rousing," Hunter explained in the memorandum, he labeled the "two distinct types of cost assumptions" "transportation" and "ammunition." "Ammunition" referred to throwaway missiles, while "transportation" meant aircraft. In sharp contrast to "ammunition" builders, Hunter pointed out, "transport aircraft operators drive their costs down to where the price of fuel is almost 50 percent of their operating costs." If orbital fuel costs became half of total costs, Hunter argued, then putting a payload in orbit would cost "only a few dollars per pound."

In the same memorandum, Hunter also attacked the belief that recoverable vehicles cost more to develop than throwaway "ammunition" because they are more complicated: "Yet we build highly complex aircraft, and man, woman and child rate them for development costs low in space budget terms. The key to low airplane development costs would seem to be the aircraft's ability to abort at any time during flight with all sorts of things wrong and successfully return to base. This, combined with a great deal of redundancy (including operational procedures) means that even with new, radical, high performance aircraft, total loss of the airplane during flight test is quite rare. There is no reason why an abortable, recoverable rocket should not be flown that way from the first flight."[33]

In the same October 1965 memorandum, Hunter also laid out a general concept for a low-cost reusable launcher. Instead of shedding rocket engines, the low-cost launcher would throw off fuel tanks. The *Atlas* and other forms of "ammunition" used engines that were heavy and suitable for either low or high altitude, but not both. Because tanks were light, engineers designed missiles that cast off engines, not tanks.[34] "The key to low cost is, in general, to reuse everything, but there are exceptions to that rule in the world," Hunter explained in a 1972 manuscript.

> If one were to throw away only fuel tanks, one might be able to set up a highly automated fuel tank production system and reduce the cost of that throwaway equipment to an acceptable value. By carrying engines, avionics and everything else in the recoverable vehicle, one could then recover all the expensive items. In other words, a stage and one-half rocket could combine all the best features of the transportation and ammunition business; and although its ultimate operating cost would perhaps never be as low as could be achieved with an all-recoverable system, it might become acceptably low enough that the difference would

not be great. Hence, the stage and one-half system represented the different idea to restimulate thinking.[35]

In his October 1965 memorandum, Hunter noted advances in engine technology. Engines now were lighter and capable of operating at both low and high altitudes. "The gambit then," he explained, "is to build the inverse of *Atlas,* and shed the tanks." "If one is throwing away only tanks," he continued, "it should be possible to build, install, and shed them using good ammunition techniques, while flying the rest of the vehicle in a good transportation mode." The resulting vehicle in its "simplest configuration" was "a rocket with auxiliary tanks, without the complication of air-breathing and without the weight and duplication inherent in two-stage rockets."[36]

The vehicle Hunter described in this October 1965 memorandum was a lifting-body craft with external fuel tanks. It was the starting point for Lockheed's *Star Clipper,* the firm's entry in the NASA shuttle competition, and prior to that, its shuttlecraft concept for the air force's Flight Dynamics Laboratory Integral Launch and Reentry Vehicle System. At this point, Hunter started and ran Lockheed's shuttle program designed around the stage-and-a-half *Star Clipper.* He was convinced that such a vehicle could lower launch costs dramatically by reducing development and operational costs. A few years later, in 1970, he told an audience at the University of Michigan that a schedule of ninety-five *Clipper* flights per year would bring per-flight costs to about $350,000, or $7 a pound of payload delivered to orbit.[37]

The *X-Rocket*

During the 1970s, Hunter became more involved in the uses of space, such as the Hubble Space Telescope and applications of high-energy lasers, rather than launchers.[38] In 1985, shortly before retiring, and even before the *Challenger* tragedy, Hunter decided it was time to start thinking about a replacement for the space shuttle.[39] He began work on a new single-stage-to- orbit launcher concept that he called the *X-Rocket.* Hunter had considered resurrecting the *Star Clipper,* but he "believed that it would be met with the usual put-down," meaning the design already had been considered and rejected. "Besides," Hunter reflected, "I didn't want to be accused of not having any new thoughts in fifteen years."[40]

He hoped that the Lockheed name might win converts in the federal government and, as a result, interest investors in the *X-Rocket.*[41] Hunter intended the *X-Rocket* name to signify that it was part of the long tradition of experimental "X" aircraft, such as the X-1 and the X-15. Alternately, Hunter called the program the XOP, for Xperimental Operational Program, "to emphasize," in his words, "that it wasn't so much the new device as the different operation which would produce the improvements."[42] Thus, the names XOP and *X-Rocket* succinctly summed up the essence of Hunter's vision: a reusable single-stage-to-orbit experimental rocket endowed with aircraft-like operations.

The *X-Rocket* represented the culmination of Hunter's thinking about applying the ways of the aircraft world to the rocket world. It paralleled the thinking of Andy Hepler and others designing single-stage-to-orbit transport within the military-industrial complex. The *X-Rocket* was a gumdrop-shaped, reusable, single-stage-to-orbit vehicle that would take off and land vertically. It could operate with or without a human onboard, "no big deal either way," in Hunter's words.[43] An aerospike engine burning liquid hydrogen and oxygen would propel the *X-Rocket.* Hunter recommended a program of up to twenty-four months to refine and validate the aerospike engine, which Rocketdyne had been developing. (The aerospike engine was a major technological breakthrough that was not yet fully available. Here was another example of a technological advance inspiring a single-stage-to-orbit concept.) From the NASP program would come the vehicle's advanced lightweight structural materials. The *X-Rocket* also would have "intact abort," meaning that with one or two of its engines out, it could still land.

In his *X-Rocket* presentations, Hunter explained the critical role of computers. Although the "commander" would be human, the "stick and rudder man" would be a computer. Electronics also would be critical, especially low-power microelectronics and precision avionics. The rocket's advanced operations electronics included monitoring and control systems and autonomous vehicle and systems checkout, operation, and fault isolation. These were under development in the Advanced Launch System, a joint undertaking of NASA and the Defense Department to build a large, expendable rocket, and various satellite programs.

The key objective of the *X-Rocket* and XOP, however, would be to eliminate "the standing army problem." "Eliminating the standing army—from day one—is the truly crucial item. It may be the only crucial item," one of his briefing charts read. NASA reported that maintaining the shuttle required a force of over 9,000 workers. Unless the problem was resolved, Hunter warned, "new vehicles of any design will make no difference." Commercial airlines required only 149 workers per aircraft, including those involved in ticketing, baggage handling, and security. Only 30 of those employees actually performed maintenance work.

The program Hunter envisioned to develop the *X-Rocket* would feature "streamlined management" and would fly the vehicle in four years. The total program cost would be $655 million, with up to $15 million for technology development over two years. This was the "faster, cheaper, smaller" way of managing a project. In addition, Hunter specifically targeted development of the *X-Rocket* to the deployment of the Strategic Defense Initiative. His briefings showed that the *X-Rocket* would be ready in time for the deployment of the first SDI sensors in 1995, followed in 2001 by the weapons deployment. As a member of the Gang of Four, Hunter was not unfamiliar with the SDI's potential launch needs.

In addition to linking the *X-Rocket* and XOP to the militarization of space, Hunter also joined it to the other half of the conservative space agenda, the commercial exploitation of space. In August 1985, he wrote a short paper, revised in April 1987 in the wake of the *Challenger* tragedy, on the need to replace the shuttle with a less expensive launch system. That paper, titled "The Opportunity," mainly outlined the *X-Rocket* and XOP as set forth in his briefings, but it also argued the need to open space to commercial exploitation. The key to conducting business in space, he contended, was a low-cost launcher like the *X-Rocket*.[44]

Hunter anticipated opposition to the *X-Rocket* from various sources. Potential private investors would want an evaluation of the concept. "The greatest experts, at least in the investors' eyes, will reside at NASA." NASA would reject the idea outright. Academics and scientists also would reject it. "Massive ridicule can be predicted." Additional opponents, he warned, would come from the "authoritative group . . . promoting air breathers."[45]

Hunter was referring to the supporters of the National Aero-Space Plane. They, he wrote, "claim that only by breathing air and using horizontal takeoff" can one achieve low orbital costs."They do not understand how good rockets can be and actually, rather hate them. They should." In the end, he concluded, "this barrier of pseudo-technical opinion" had to be surmounted and a rocket-powered vehicle accepted.[46]

Hunter faced further opposition from his own company, the Lockheed Missiles and Space Company. He used his position to compel the firm's Advanced Development Division to analyze the *X-Rocket.* The vehicle they studied used a cluster of upgraded RL-10 *(Centaur)* rocket engines, not aerospike engines. The Advanced Development Division confirmed Hunter's basic concept. However, after he began promoting the *X-Rocket* outside the firm to government officials, Lockheed ordered a second review by its Missile Systems Division, whose sole business was the design and manufacture of ballistic missiles launched from Trident submarines. Not surprisingly, their review was not favorable; they concluded that the vehicle would not be able to carry a payload. At this point, Lockheed dropped further consideration of the *X-Rocket,* and Hunter retired.[47]

Between 1985 and his retirement in 1987, Hunter tried to raise interest in his *X-Rocket* outside Lockheed. He briefed myriad military and civilian officials, including such air force organizations as the Astronautics Laboratory, Edwards AFB; Phillips Laboratory; the Office of the Secretary of the Air Force; and the Space Division. Ivan Bekey, then a NASA consultant, expressed interest in the *X-Rocket.* He arranged for Hunter to brief Richard Truly, then NASA associate administrator for spaceflight, in December 1986, and Dale Myers, NASA deputy administrator, in February 1987. Hunter also gave a series of presentations to officials with the Strategic Defense Initiative Organization, starting with the Innovative Architecture Blue Team under Frank Hammill, and eventually reaching Gen. James Abrahamson, then SDIO director, in a briefing during the summer of 1987.[48]

After leaving Lockheed, Hunter gave the *X-Rocket* and XOP a new name, the *Single Stage eXperiment* or *SSX* (which later stood for *Space Ship eXperimental*).[49] He described the *SSX* in a new paper, "The *SSX Single Stage Experimental* Rocket," written March 15, 1988.[50] A single-stage-to-orbit vehicle was now possible, Hunter wrote, because the NASP program was developing the novel lightweight alloys and composite materials needed for fabricating the aeroshell and fuel tanks. This was the same technology argument traditionally used to justify a new single-stage-to-orbit concept. The *SSX* also would be able to abort and would be developed in an incremental flight test program like an airplane. "The first flight will likely rise only a few hundred feet, hover for only tens of seconds, then land. The test program will work up the flight envelope as airplanes do, will have practiced abort procedures on all flights, and will not proceed to orbit until the vehicle is darn good and ready."[51]

Above all else, the key to the *SSX,* like the *X-Rocket,* was operations (operating rockets like airplanes, meaning autonomous operations), a minimal launch crew, and a short turnaround time. A computer would do the actual flying of the *SSX,* not a human, though the *SSX* "*commander* remains a *human,*" Hunter explained. The *SSX* would require a far smaller ground crew than that required for the shuttle, and would be more comparable to that required for commercial or military aircraft. In the *SSX* paper, Hunter also specified that a small team with a single line of authority and adequate funding would design and build the *SSX.*[52] These were characteristics of a "faster, cheaper, smaller" program.

The *SSX,* then, and its *X-Rocket* and XOP antecedents, started from customary single-stage-to-orbit thinking that found inspiration in new technologies, especially rocket engines. The *SSX* was no different, in that novel lightweight alloys and composite materials for manufacturing the aeroshell and fuel tanks were its technological inspiration. Hunter broke with that mold, however, and added aircraft operations, by which he meant abort capability, autonomous operations, a minimal launch and maintenance crew, and a short turnaround time. The *SSX* went another step farther by adding a specific managerial approach ("faster, smaller, cheaper"). The combination of single-stage-to-orbit "X" vehicle, aircraft-like operations, and "faster, smaller, cheaper" program management was unique and new, and certainly distinguished it from the typical long-term, expensive, colossal programs of the cold war carried out by the military-industrial complex. Hunter's *SSX* vision, though, would have remained just that, a vision, like the single-stage-to-orbit designs envisioned by Gary Hudson, Phil Bono, and Ed Gomersall, if it had not been for Hunter's attending a meeting of the Citizens' Advisory Council on National Space Policy in 1989.

Part III / The *Space Ship eXperimental*

CHAPTER 6

Launching the *SSX*

Not one, but a series of critical events allowed Hunter's vision of the *X-Rocket* to become a real experimental spaceship. The first took place at a meeting of the Citizens' Advisory Council on National Space Policy. It was one of many diverse citizen space interest groups that formed what might be called the space movement. The nature of these groups underwent a significant and lasting change around 1980, as the movement became political for the first time and as science fiction spurred fresh interest in space. This change also coincided with the election of Ronald Reagan and the ascendance of the conservative space agenda. The change in the space movement paralleled the policy course set in the White House: new space groups promoted the conservative space agenda. Many of these groups supported the Strategic Defense Initiative, while others promoted the commercialization of space. Key figures in the politicization of the space movement also were leading proponents of the conservative space agenda, such as Jerry Pournelle, Lt. Gen. Daniel Graham, and Republican activists associated with Newt Gingrich.

The Space Movement

The oldest citizen space interest groups tended to be devoted primarily to education and public information, although some, such as the Space Studies Institute, conducted internal research and funded external research. During the 1970s, following the end of the Apollo program, popular support for NASA's space program diminished. According to an annual survey conducted by the National Opinion Research Center at the University of Chicago, the space agency's popularity declined from its height in 1969 to an all-time low in 1975. In that year some 60 percent of those polled felt that the amount of money being spent on space was too much, while only 7.4 percent felt the amount was too little. Beginning in 1976, though, support for the space program started to rise. The increase did not result from any NASA success or presidential programs, but reflected a rising general interest in space fiction.

Feeding this interest were such hit movies as *Star Wars, Close Encounters of the Third Kind,* and *Star Trek;* the television series *Battlestar Galactica* and *Buck Rogers;* and *Omni* magazine. All made their debuts between 1978 and 1982, which coincidentally were boom years for pro-space groups.[1]

About this same time, the space movement underwent a dramatic change as a number of groups formed in order to support political activities. Certain government agencies, notably the air force and NASA, already promoted space programs, and some professional organizations, such as the American Institute of Aeronautics and Astronautics, and the aerospace industry lobbied as well. However, these were far from constituting a space movement that rallied citizens in support of space programs.

Leading the trend toward political activism was the L-5 Society, founded in 1975.[2] It promoted space exploration, space stations, and space colonies; published the *L5 News* newsletter; and maintained a nationwide space activist telephone network. It also held annual meetings and supported many regional and local activities. The link connecting the L-5 Society with the conservative space agenda was Jerry Pournelle, founder of the Citizens' Advisory Council on National Space Policy. He was a longtime L-5 board member and at one time its de facto co-president.[3]

As early as 1977, taking advantage of new Internal Revenue Service regulations that allowed educational organizations to spend up to 20 percent of their income on lobbying, the L-5 Society launched its Legislative Information Service to influence Congress. The society hired a professional lobbyist in August 1979 specifically to fight against a United Nations treaty dealing with the Moon (and won), and later to influence congressional legislation and to raise funds from industry.[4] This was only a first step in the politicization of the space movement. Other groups formed specifically to promote the conservative space agenda.

The early 1980s saw the rise of pro-military space interest groups whose specific goal was to support the Strategic Defense Initiative. By 1985, the fastest membership growth in space interest groups took place in those dedicated to political activities, specifically, the conservative space agenda. Two organizations alone accounted for 95 percent of the growth in membership: High Frontier and the American Space Foundation.[5] The growth of those and other similarly inclined groups reflected an increasing shift among conservatives toward support for space programs, including large-scale projects, usurping the position held by liberals during the 1960s.

The American Space Foundation (ASF) grew out of a December 1980 conversation between Congressman Newt Gingrich and Robert Weed, a young but experienced Republican political activist who had worked on Gingrich's campaign. Acting on a suggestion from Gingrich, Weed set up a political action committee (PAC) for space called the American Space Political Action Committee. He brought together other young Republican political workers he knew, including several from Gingrich's staff, to raise funds for pro-space candidates and to fill what they perceived to be a vacuum in U.S. space policy. When the American Space PAC failed to attract enough support, its creators reorganized it as the American Space Foundation, a lobbying organization that used a direct-mail approach to raise funds.

In January 1983, ASF began publishing a quarterly newsletter, *ASF News.* Later that year, the foundation set up an office in a building on Capitol Hill owned by the conservative Heritage Foundation (which also supported Daniel Graham and High Frontier). In 1984, the ASF lobbied for a program that echoed the conservative space program espoused by Newt Gingrich. They called for the establishment of a permanently manned space station; the increase of NASA funding to a level of one percent of the federal budget; the fostering of private sector development of space; and the promotion of a space-based defense system to prevent nuclear war, namely, the Strategic Defense Initiative. The Reagan administration supported all but the second of these objectives.[6]

High Frontier, founded in 1981, was the creation of Daniel Graham. It grew out of a defense study that Graham organized shortly after Reagan's election. He and Robert Richardson, a retired air force general, assembled a study team under the auspices of the conservative American Security Council Foundation's Wedemeyer Strategy Center, later in association with the Heritage Foundation. Among the space advocates involved in the study in one way or another was Jerry Pournelle, an officer of the L-5 Society and chairman of the Citizens' Advisory Council on National Space Policy.[7]

In March 1982, the Heritage Foundation published the group's report as *High Frontier: A New National Strategy.* It advocated a space-based missile defense called the global ballistic missile defense (GBMD) system.[8] The explicit aim of the High Frontier Society was to promote the ballistic missile defense plan as a purely defensive system. High Frontier began publishing a newsletter in 1983, and by 1984, the group had grown to be one of the largest American space interest groups with 40,000 members and an annual budget of $700,000. High Frontier spun off a political action committee called the

American Space Frontier Committee, which was successful in raising money to support the congressional campaigns of pro-SDI candidates.[9]

The Citizens' Advisory Council

The Citizens' Advisory Council on National Space Policy was yet another of the many and diverse citizen space interest groups formed around 1980. The council also reflected the shift toward political activism that took place within the space movement at that time. It favored both the Strategic Defense Initiative and the commercialization of space. Like the space movement in general, the Citizens' Advisory Council had close ties with both the science fiction community and the military-industrial complex. However, it was different in that its reason for being originated in the White House. The group arose and prospered during the Reagan years and met only once after Reagan left office. Toward the end of the Reagan presidency, during a meeting of the council, Max Hunter would find a patron for his *SSX*, Daniel Graham.

The founding of the Citizens' Advisory Council had its roots in Ronald Reagan's 1980 election victory. Reagan's transition team asked Gen. Bernard Schriever, who had overseen the creation of the United States' missile programs during the 1950s and 1960s, to write the space and space defense sections of the transition team papers. Schriever turned to Francis X. Kane, a retired air force colonel who had worked with Pournelle on previous occasions, and Kane asked Pournelle to help.[10] As for its role in the Reagan transition team, Pournelle described the Citizens' Advisory Council in 1989 as "an entity created as part of the Reagan transition team structure, but kept around because Mr. Reagan and some of his people found our reports useful."[11]

The first meeting took place over the weekend of January 30–February 1, 1981, in the Tarzana, California, home of science fiction writer Larry Niven. Mrs. Marilyn Niven managed and catered the gathering. Organizing and chairing the meeting was Jerry Pournelle, who often penned science fiction novels with his friend, Larry Niven. His connections to the incoming Reagan administration provided Pournelle the credibility to hold a large meeting on space policy in the Nivens' home. "After all," Pournelle later explained, "major players in the space and defense community were not likely to pay their own way to a meeting called by a couple of science fiction writers even though Harry Stine [a science fiction writer] and I both had professional employment in the aerospace industry."[12]

Because the invitation to attend the meeting specifically requested assistance in preparing the Reagan transition team papers on space, Pournelle was able to induce a range of representatives of the military-industrial complex to attend. They included astronaut Buzz Aldrin; George Merrick, head of the North American Rockwell shuttle program; Fred Haise, then president of Grumman; Brig. Gen. Stewart Meyer, formerly commander of the Redstone Arsenal; Stewart Nozette, soon to be named to the National Space Council; space enthusiast and lawyer Arthur Dula; Lowell Wood, from Lawrence Livermore; and Gordon Woodcock, Boeing engineer and president of the L-5 Society in 1984. Max Hunter, Gary Hudson, and Daniel Graham also were among the attendees.

Subsequent meetings also featured representatives of NASA and the military-industrial complex. From the space agency, for example, came NASA Administrator Thomas O. Paine and astronauts Gerald Carr, Philip K. Chapman, Gordon Cooper, Walter Schirra, and Charles "Pete" Conrad. Among the retired military officers attending were Brig. Gen. Robert Richardson, U.S. Air Force; Brig. Gen. Stewart Meyer, U.S. Army; Col. Jack Coakley, U.S. Army; and Col. Francis X. Kane, U.S. Air Force. Konrad Dannenberg, a former member of von Braun's rocket team, frequented meetings, as well.

Pournelle also invited a number of science fiction writers with technical backgrounds, such as Robert Heinlein, Poul Anderson, Dean Ing, and G. Harry Stine.[13] In attendance too was Betty Jo ("Bjo") McCarthy Trimble, who had headed the letter-writing campaign to keep Gene Roddenberry's *Star Trek* television series on the air in 1968, and who had led another letter campaign to name the first shuttle orbiter the *Enterprise.*[14] It was not unusual for science fiction writers to participate in space interest groups. Ben Bova, for example, headed the National Space Institute, founded by Wernher von Braun in 1974, and, in general, science fiction fans and scientists constituted the great majority of space interest group membership.[15]

According to Pournelle, the Reagan transition team integrated the Citizens' Advisory Council report from that first meeting into its papers provided to the Reagan administration, and a copy of the report also went directly to Richard Allen, Reagan's first national security adviser.[16] Allen placed a copy of the report on Reagan's desk, and the president read the entire report. Allen informed his successor about the Citizens' Advisory Council, and Reagan read and acknowledged council reports "during his whole term of office." As evidence, Pournelle cites that: "Phrases from our reports were incorporated into

Mr. Reagan's SDI speech of March 23, 1983."[17] Victor Reis, then assistant director to Reagan's science adviser, George Keyworth, believed that the council's reports did not reach the president and had little impact on policy. However, a comparison of Reagan's SDI speech and council reports undertaken by Michael A. G. Michaud shows some similarities,[18] suggesting that the reports may have had an impact on Reagan policy speeches. In any case, the council did not meet between 1988 (the end of the Reagan presidency) and the August 10, 1997, meeting that NASA Administrator Dan Goldin attended.[19]

The issues discussed in Citizens' Advisory Council meetings and reports closely paralleled space policy issues that the Reagan administration deliberated. Moreover, the thinking of the council closely reflected the space ideas of Newt Gingrich and the conservative space agenda. Like Graham and Hunter, council members saw the military and commercial uses of space as linked, with the need for low-cost access to space as a necessary first step. In its discussions of low-cost launchers, the Citizens' Advisory Council rejected the space shuttle because of its high operating costs. Following the *Challenger* disaster, the council, like President Reagan, continued to express its support for the shuttle as well as the space station.

Nonetheless, the *Challenger* incident did provide the council an opportunity to do more than put together the usual white paper and lobby the White House. With funding from the L-5 Society and a private foundation, the group held an extraordinary three-day meeting in Larry Niven's home May 9-11, 1986. Its purpose was to consider solutions to the crisis in space transport that resulted from the *Challenger* incident. The outcome of that meeting was an eighty-four–page white paper, *America: A Spacefaring Nation Again,* edited by council members Henry Vanderbilt (later founder of the Space Access Society) and G. Harry Stine.[20] That white paper encapsulated much of the Citizens' Advisory Council's thinking and echoed the space views of Newt Gingrich, the Reagan administration, and the conservative space agenda.

A Spacefaring Nation

Like many of President Reagan's pronouncements on space policy, *America: A Spacefaring Nation Again* evoked the memory of President Kennedy. The council's stated goal was to make America "a spacefaring nation," in fulfillment of President Kennedy's 1962 pledge. "Spacefaring nations," the report explained, "have space navies, space factories, space miners exploiting space resources, space hotels, and space tourists. . . . A real spacefaring nation would have a Hilton on the Moon, and a lively debate about whether to allow a McDonald's to open within a mile of Neil Armstrong's first footprint." The white paper endorsed continued operation of the space shuttle and construction of the "Big Space Station," the same position taken by the Reagan administration.

The council report argued for routine economical access to space in order to achieve the three goals of space exploration, defense, and commercial exploitation.[21] It was specific about how to realize these goals. Moreover, in elucidating a program for achieving low-cost routine space launch, the report echoed the thinking of Max Hunter and his *X-Rocket (SSX)*. The first step was not to do "business as usual" but to implement a "fast track" management style. By way of example, the report cited Hunter's management of the *Thor* missile project from contract to operational rocket in under three years. Other examples were the development of the SR-71 and the *Polaris* both in less than four years.

The *Spacefaring* white paper also echoed Hunter's call to use "X" vehicles analogous to the X-1 and the X-15. Although the National Aero-Space Plane was an example of one such "X" program, the report urged: "We should not stop with just one such program. We should immediately pursue parallel programs to explore single-stage-to-orbit rocket vehicles and other potential military launch vehicle designs as well."[22] Here were three elements of Hunter's vision developing through council discussions: a "rapid" management approach, the use of "X" vehicles, and single-stage-to-orbit transport. Clearly, the council was aware of, and agreed with, Hunter's thinking before he addressed the group in December 1988. These ideas, moreover, were coalescing into a single vision within the context of the conservative space agenda.

The *Spacefaring* report explicitly supported the Strategic Defense Initiative (SDI) and the commercialization of space. For example, it called for the immediate deployment of certain SDI components, such as the "chemical-powered laser space battle stations" (the work that Hunter had overseen at Lockheed) and "'space shotguns' firing projectiles which destroy ballistic missiles with the kinetic energy of impact."[23] It also urged passage of the Commercial Space Incentive Act, intended "to stimulate private enterprise to develop low-cost space-launch capabilities without having to compete with government space launch systems." The Citizens' Advisory Council member who wrote the Commercial Space Incentive Act section was Jim Muncy.

Muncy, as we saw, was a space activist associated with Newt Gingrich and President Reagan's science advisor, George Keyworth. In 1988, he founded the Space Frontier Foundation.[24]

The essence of the Commercial Space Incentive Act was a payment guarantee by the federal government of $500 per pound for the first one million pounds of payload placed in orbit by private American companies each year for a period of ten years. The act offered an additional incentive of $250 per pound if the launcher carried humans into orbit. The minimum payload weight to qualify for the incentive payment was 10,000 pounds. The act also would require all federal agencies to place their payloads on private launchers, using government launchers only when private rockets were not available. The Defense Department would be exempt from this provision in cases when "suitable safeguards relating to national security cannot be maintained."[25]

The purpose of the act was not to put payloads in orbit, but to provide an incentive payment to encourage the development of low-cost launchers able to put payloads in orbit for under $500 per pound. The model for the Commercial Space Incentive Act was the Pacific Railroad Act of 1862 and the Kelly Bill of 1925, which sought (in the language of the bill) to "encourage commercial aviation and authorize the Postmaster General to contract for airmail service." As the council report proclaimed: "The entire United States air transportation industry—the largest in the world—owes its existence to the Kelly Bill."[26] Ultimately, the Citizens' Advisory Council wanted the government, in particular NASA, out of the business of launching payloads and wanted private enterprise to take over the task. The Commercial Space Incentive Act never became law. In fact, it never survived beyond the pages of the Advisory Council's report, and it certainly was never heard of in the nation's capital. Nonetheless, it was in tune with the thinking of the Reagan administration and the conservative space agenda.

The *Spacefaring* report of the Citizens' Advisory Council received wide circulation. A copy went to the White House, along with a cover letter signed by Jerry Pournelle and addressed to President Reagan. The L-5 Society (where Pournelle had influence) published and distributed the report to its members.[27] Following the release of *America: A Spacefaring Nation Again,* which urged government support of "single-stage-to-orbit rocket vehicles and other potential military launch vehicle designs,"[28] the Citizens' Advisory Council turned its attention to finding, in the words of member Harry Stine, "a specific spaceship type that could be suggested for immediate government backing."[29]

Launching the *SSX*

Consequently, a Citizens' Advisory Council meeting took place over the weekend of December 5, 1988,[30] shortly after the election that brought George H. Bush and J. Danforth Quayle to the White House. Quayle soon would be chairing the revived National Space Council, which coordinated space policy among federal agencies. Pournelle and the Citizens' Council had hopes of influencing Quayle and the National Space Council.[31] Pournelle's plan for the meeting was to have Max Hunter deliver a presentation on the *SSX,* followed by a presentation on the *Phoenix* by Gary Hudson. Then he would ask those present for suggestions.

A few weeks prior to that meeting, Hunter and Hudson visited Pournelle and showed him a briefing they had prepared on why the air-breathing National Aero-Space Plane (NASP) was the wrong approach. Rocket engines, not air-breathing scramjets, were the best propulsion for a single-stage-to-orbit transport, they argued. As Pournelle explained later, "many, myself included, remained wedded to the idea of air breather first stages, or air breathers that converted to pure rockets after going at very high speeds in the atmosphere." Indeed, in his novel *Fallen Angels,* published in 1991 with co-authors Larry Niven and Michael Flynn, Pournelle portrayed astronauts flying through the atmosphere and scooping up air in vehicles equipped with engines reminiscent of NASP.[32] "By the time [Hunter and Hudson] finished," Pournelle wrote, "I was convinced that NASP was the wrong way to go, and that we ought to start over with a simple pure rocket."[33] As a result, Pournelle joined Hudson and Hunter in supporting a rocket-powered, single-stage-to-orbit launcher.

The accounts of what happened during the December 1988 Citizens' Advisory Council meeting are as varied as those making the account. Max Hunter, for example, recounted: "Pournelle asked me to go first (he should have known better!). I unloaded the full SSX story complete with four-letter words." Pournelle then asked the entire group for other views. When none were forthcoming, "he specifically asked [Gordon] Woodcock [from Boeing, the only employee of a major aerospace company at the meeting] his opinion and Gordie replied he thought it was the way to go. Pournelle then asked for other suggestions and the room was dead silent. Jerry was visibly shaken by this. He had expected many suggestions and that the group would then try to sort them out over the weekend. Design thinking had terminated after the first presentation."[34]

Pournelle expected opposition to Hunter's rocket-powered, single-stage-to-orbit proposal from some of the air force people unofficially present (believers in air-breathing engines), but they expressed none.[35] Graham, also at the meeting, reflected in amazement: "Somewhat to my surprise, no one seriously challenged Hunter's technological assertions."[36] Gary Hudson rendered a less dramatic account: "The meeting highlight was a presentation on the vehicle known as SSX, or Space Ship Experimental. (Hunter had renamed the *X-Rocket* following his departure from Lockheed.) Besides Hunter and myself, Daniel O. Graham was in attendance. He and the rest of the council agreed to endorse SSX."[37] Council member Harry Stine, on the other hand, recalled the Hunter and Hudson briefings as taking place on two separate occasions: "At the first 1988 Council meeting, Hunter presented a briefing to the Council on an SSTO [single-stage-to-orbit] concept called SSX . . . After much discussion and another weekend meeting, the Council members supported this evolution of the SSTOs of Gary Hudson and Max Hunter. A white paper and supporting technical documentation were prepared."[38]

Daniel Graham cast the council meeting in an entirely different light. He recalled that Graham and Hunter had met well before the meeting of the Citizens' Advisory Council because of their mutual interest in SDI, as well as their disagreement over its implementation. Hunter championed a laser-based defense, while Graham favored kinetic-kill weapons. As Graham reminisced in his autobiography: "For years we [High Frontier] had sought ways to lower launch costs, because according to solid estimates from DoD [Department of Defense] of the costs of deploying the spaceborne elements of SDI, a large percentage were launch costs. In the fall of 1988, I learned of a solution. Maxwell Hunter, a well-respected and able aerospace engineer, came to my offices."[39]

When Hunter explained his *SSX* idea to Graham, Graham saw the *SSX* as having implications for all space needs: "I was highly impressed by Hunter's proposal. Here, at last, was a solution that involved a fundamental improvement in space transportation of revolutionary potential for enhancing military, civil, and especially *commercial* uses of space." Graham later related: "I was excited about the SSTO idea, but I knew my technological limitations and was not ready to throw this cat into the space community canary cage without hearing from other experts. I suggested to Max that we call together a group of such experts and get their reactions. He agreed, and I contacted Dr. Jerry

Pournelle to call together his loose confederation of experts and enthusiasts, the Citizens' Advisory Council on National Space Policy, to critique Max's proposal."[40]

The varying and contradictory versions of what transpired at the December 1988 meeting of the Citizens' Advisory Council, like the sketchy and conflicting recollections portrayed in the film *Rashomon,* leave one wondering exactly what happened and when. A clue to the content of Hunter's presentation to the Citizens' Advisory Council is his paper, "The SSX Single Stage Experimental Rocket," written March 15, 1988, in which he laid out the same ideas briefed to the council later that year.[41] For the December 1988 meeting, Hunter renamed it the *Space Ship eXperimental,* to indicate that it was an experimental program to demonstrate operational concepts.[42] It contained the same set of ideas laid out in his earlier *X-Rocket* and XOP briefings.

As with the single-stage-to-orbit concepts of Phil Bono, Ed Gomersall, and others, a technological advance made the *SSX* feasible. In this case, it was the availability of new lightweight alloys and composite materials for the aeroshell and fuel tanks. These were the fruit of the NASP program. To this, Hunter added aircraft-like operations: autonomous operations, a minimal launch crew, short turnaround time, and the ability to abort. "The real difference [between rockets and airplanes]," he wrote, "is the intact abort capability of the airplanes, especially in the event of engine failure." The government would develop the *SSX* through an incremental flight test program using an "X" vehicle. Computers would play a pivotal role in the *SSX,* because a computer would do the actual flying of the vehicle. A small team with a single line of authority and "adequate funding" would design and build the *SSX.*[43] Thus, not only the government agency, but also the contractor would operate a "faster, cheaper, smaller" program.

Because the *SSX* combined a single-stage-to-orbit concept with aircraft-like operations, use of an "X" vehicle, and "faster, cheaper, smaller," it was unlike any other single-stage-to-orbit project previously proposed. It certainly would not be another long-term, expensive, colossal, single-stage-to-orbit program like NASP run in the manner of a typical cold war program carried out by the military-industrial complex. Hunter's presentation of the *SSX* conceptual package won over the attending members of the Citizens' Advisory Council. One of the members present at the meeting, Daniel Graham, made Hunter's *SSX* the subject of a personal crusade.

According to Hunter: "The principal action was that Dan Graham decided that High Frontier would work to support this and volunteered to use his good offices to get time with the vice president. . . . This also started a much closer working relation between Graham and myself."[44] Graham expressed it simply: "I expended my political blue chips with Dan Quayle to get an appointment to sell him the SSTO idea."[45] Now, Hunter's *SSX* had not only the backing of the Citizens' Advisory Council (an unclear advantage in and of itself, given Hunter's earlier failure with the Lockheed name behind his efforts), but also the political support of Graham and his High Frontier national organization. As Graham pledged at the meeting, he would unlock the door to the White House. Once in, though, what would they say to the vice president?

A Presentation Strategy

Because of the high importance of the Quayle briefing, a great deal of effort went into preparing it. At one point, as understood by Max Hunter, five people would make the presentation, each with his own objective. After Graham introduced the participants, Pournelle would talk about the importance of the *SSX*, "the big picture." Next, Hunter would provide a technical description of the project and background on "fast" programs. General Schriever then would discuss "fast" programs by providing examples from his personal experience, and Lt. Gen. Thomas Stafford simply would provide a personal endorsement of the concept and program. Finally, Graham would furnish a summary and "offers of undying support, pleading, final flag wave."[46]

The goal of the meeting with Quayle was "to try and get some form of administrative support for the pursuit of the SSX concept." In order to accomplish this goal, Graham, Hunter, Pournelle, and friends realized that they needed to "present a pursuasive [*sic*] briefing on behalf of the proposed program," to "know exactly what we would like the V.P. (or his Space Council) to do if he accepts our arguments," and to "recognize, and be prepared to deal with, the fact that to date responsible government agencies and Committees have ignored the SSX/Phoenix approach in their studies." Setting up the briefing was not hard, and how to respond to why "the SSX/Phoenix approach" had been ignored was clear: "In brief it is truly a NEW APPROACH." However, "Exactly what we should recommend to the VP is less clear."[47] The references to the "SSX/Phoenix approach" are confusing in retrospect, insofar as the *SSX* was Max Hunter's design, while the *Phoenix* was that of Gary Hudson.

One option was to have the program funded and managed by the SDIO. But the SDIO option was fraught with political difficulties. "We are an 'outside' group and there has been much criticism of the Reagan administration over SDI and especially the fact that the president did not bring in all the responsible officials for advice BEFORE initiating a program," one member of the group wrote. "Whether right or wrong, and justifiable or not, this criticism will surely not [have] been lost on the Bush/Quayle crowd." Therefore, before Quayle could ask the SDIO to undertake a program they had not proposed themselves, he would have to consult with "some IN HOUSE people or group. If this is near inevitable," the strategy piece ran, "then we should figure out how to make sure that this 'group' will be supportive to objective on SSX and not automatically antagonistic. This in turn suggests we might wish to recommend it up front instead of letting other[s] do this.?? [*sic*]"[48]

The SDIO, nonetheless, *was* the logical choice for the *SSX*'s institutional home. According to Hunter, "SDIO already had a reputation for a light-handed management style that got results, and, Graham believed, the agency was too young for 'bureaucrats' to hinder the project."[49] Ironically, Hunter had briefed the former head of the SDIO, Lt. Gen. James Abrahamson, on the *X-Rocket* in the summer of 1987. Abrahamson had serious concerns about the engine configuration, specifically, that there were too many engines to produce a reliable system. He was especially concerned about what he thought were miniature turbopumps. However, he did not express these concerns during the briefing. What Abrahamson thought were individual engines were actually individual thrust chambers fed by a manifold system from eight sets of turbopumps. This misconception did not clear up, though, until after Abrahamson's retirement, when he stated (without elaborating) that if he "had understood this, things would have been different."[50] However, it is not clear what Abrahamson meant by "things" or "different."

Graham, Hunter, Pournelle, and the others considered several agencies as potential homes for the *SSX*. According to Graham: "The reasoning, with which I fully agreed, was that the other likely candidates for housing SSTO, NASA and Air Force, would stifle our baby in its crib. In NASA it would be viewed as a threat to hoped-for prolongation of the Shuttle program; in Air Force it would be viewed as a threat to the hoped-for Advanced Launch System (ALS), a follow-on expendable rocket system. Further, SDIO already had a reputation for a light-handed management style that got results. It was too young an agency to provide dozens of bureaucrats to interfere with the worker bees."[51] Another potential home for the *SSX* was the Advanced Research Projects Agency (ARPA),[52] which let contracts for a good deal of Strategic Defense Initiative research. Given Graham's strong support of SDI, and the eventual presentation of the *SSX* as a way of implementing the SDI Brilliant Pebbles concept, it was more logical that the SDIO run the *SSX* program.

Graham recalled that Vice President Quayle asked which agency should run the program. "He asked whether he should urge the SSTO [single-stage-to-orbit] effort on NASA, and was taken aback when I said no. He was apparently even more puzzled when I opposed the second choice, Air Force. I urged that the job be given to SDIO as the organization without enough colonels and ranking civilians to louse up the effort."[53]

A further obstacle to approval of the *SSX* would be any committee that the vice president created to look into the program, as "these and not the VP would decide what would be done and how." The vice president would not order the SDIO to support *SSX*. So what strategy would work? "Persuade the VP to appoint a Committee to look at the SSX approach and recommend what to do. If he wants his Council to decide this then: Persuade the VP to let us brief the Council. If he agrees to the Committee idea then: be prepared to suggest names to be on it and ask to have a role."[54]

The Pitch

Then came the day to meet Quayle, February 15, 1989. The three who carried out the briefing were Max Hunter, Daniel Graham, and Jerry Pournelle. Graham recalled: "I asked Max Hunter and Jerry Pournelle to accompany me. Curiously, Mr. Quayle at first thought I was there to discuss John Tower's confirmation problems. I told him, no, I thought I had a more important issue."[55] The briefing did not take long, "about 25 minutes," according to Hunter.[56] Graham began the meeting with some concise background information before outlining the basic *SSX* idea: a reusable spaceship that would dramatically reduce the cost of getting to orbit. Graham recounted: "I gave him a quick brief on the SSTO and its enormous potential and presented him with a more detailed paper on the subject."[57] Hunter followed with a more detailed discussion of the *SSX* vision.[58]

He explained that the goal was to cut launch costs by making the vehicle reusable, by reducing the number of people required to maintain and operate it, by making it capable of an aborted landing, and by shortening vehicle turn-

Fig. 2. February 15, 1989, meeting of *(from left to right)* Vice President Dan Quayle, Daniel Graham, Max Hunter, and Jerry Pournelle in the White House. (Courtesy of Steve Hoeser.)

around time. The *SSX* would be run as a "fast" program. It would cost under a billion dollars and run only four years from start to test flights. On an annual basis, the program would cost less than one shuttle flight, they told the vice president. The *SSX* would take off and land vertically, from anywhere at any time. It would have an aerospike engine with multiple (16 to 24) engine modules. Reflecting years later on his technical presentation to Quayle, Hunter wrote: "As you know, the SSX story, coming out of the blue, is a rough show for an experienced rocket engineer to absorb, so I have to believe the Vice President missed a few technical details."[59]

Graham then told the vice president what they wanted him to do. "Bear in mind that, as the name implies, we are proposing an experiment—not a full-blown program." He recommended that the "proposal be discussed with NASA's expert on this type of vehicle, Dr. Ivan Bekey [a NASA advocate of single-stage-to-orbit concepts]" and urged "that this concept be briefed to the National Space Council with a view to funding at a modest level of [*sic*] proof-of-concept prototype SSX."[60]

Before taking questions, Graham introduced Pournelle, who talked about the effects of a new space program on education and "national purpose." He discussed the political value of having supported the project when Quayle sought the presidency at a future date. "Dr. Pournelle added some thoughts," Graham reported, "based on his experience in California politics, about the effects such a development could have on elections six or eight years hence—i.e., when Dan Quayle's shot at the presidency would be pending."[61]

At the time of the briefing, Quayle had been vice president for less than a month. The *SSX* program schedule, as pitched to the new vice president, was filled with political stratagems geared toward the 1992 presidential elections. The timing of the public announcement could be made "at about the time of the State of the Union Address in 1991," Pournelle later wrote Quayle, "and will be followed by a number of safely predictable newsworthy achievements in 1991 and 1992. SSX can also perform a suitably significant mission to mark the 500th anniversary of Columbus's discovery of the New World on October 12, 1992."

As Pournelle left Quayle's office after the February 1989 briefing, he recalled Quayle asking "if this really could be done before 1992." His answer, written in a subsequent letter, was: "Yes. We can do something both significant and spectacular in space before 1992."[62] The specific program schedule presented to Quayle, according to Pournelle, called for the critical design review and final funding commitment to take place in December 1990; the public project announcement in January 1991; first flight test in spring 1992; and orbital flight on Columbus Day, 1992.[63]

Did Quayle understand the importance of the *SSX* proposal? Graham observed that: "It appeared that Mr. Quayle, although a very 'quick read,' i.e., able to grasp ideas quickly, was not much interested in the details of SSTO technology."[64] How momentous was this meeting in Quayle's mind? In an autobiographical memoir of his years as vice president, Quayle devoted a respectable number of pages to space issues. He discussed the Space Exploration Initiative (a mission to return to the Moon and to land an astronaut on Mars), the space station, and conflict with NASA, which Quayle characterized as "a very pampered bunch." However, he failed to mention the SDIO's SSTO program or the *DC-X* at all.[65]

The Bush administration did not give space the high national priority that it enjoyed under the Reagan administration, with the possible exception of the Space Exploration Initiative. Nonetheless, Quayle helped the *SSX* project by handing it over to the National Space Council for analysis. "To me," Hunter later wrote, "that was all we needed from him at that point."[66] The meeting with Quayle also resulted in the vice president's request that Graham and company brief the National Space Council.[67] Those briefings were only part of a larger effort initiated after the Quayle meeting to sell the *SSX* program around Washington.

Graham promoted the *SSX* through his High Frontier organization, which ran promotional pieces in its newsletter.[68] On March 11, 1989, Hunter revised his *SSX* paper, now called "The SSX, Space Ship eXperimental," for the campaign. However, he did not participate personally in the Washington campaign as much as Graham. As he later explained: "Since I anticipated that Graham would likely set up meetings on short notice, I told him that although I came to DC frequently, I was not in position to guarantee instant response to his every whim, so I needed an alter ego in DC. We set up Steve Hoeser of GRC [General Research Corporation, Arlington, Virginia] as that person."[69]

As early as 1983, Hoeser, then an air force officer, had become involved in the Military Aerospace Vehicle (MAV) program through the Air Force Space Division program office. MAV was a generic name given to all vehicles with single-stage-to-orbit capabilities, such as the Advanced Military Spaceflight Capability (AMSC) program, which was better known as the Trans-Atmospheric Vehicle (TAV) program. In 1989, Hoeser was with a consulting firm, the General Research Corporation, and he now would become Graham's aide-de-camp in the battle to transform the *SSX* concept into a funded government program. Hoeser first met Graham at the house of science fiction writer Larry Niven during a meeting of Pournelle's Citizens' Advisory Council. "The man really impressed me," Hoeser later recalled. He "just impressed me as a guy who really understood the political system, and how things could get done in that arena."[70] It was Graham and Hoeser, then, who carried out much of the *SSX* Washington campaign, often accompanied by members of High Frontier.

The strategy behind the Washington meetings and press conferences was to link the *SSX* to Brilliant Pebbles, a new SDI plan. Brilliant Pebbles was the brainchild of Lawrence Livermore scientists Gregory Canavan, Edward Teller, and Lowell Wood. They hoped to resolve the two major drawbacks of the space-based interceptor (SBI) that constituted one of the six major subsystems of the Strategic Defense System (SDS) Phase I architecture. First, the SBI was large and therefore an easy target for Soviet ASAT weapons. Second, it was very expensive. The solution to these problems was to use advances in miniaturized sensors and computers to develop an interceptor that could operate without the sensors and communications equipment of the SBI garage. Brilliant Pebbles was that solution.

Brilliant Pebbles consisted of a large number of lightweight, low-cost, single hit-to-kill vehicles that provided integrated sensors, guidance, control, and battle management. Basically, Brilliant Pebbles would be an autonomous space-based defensive interceptor. Each Pebble would have its own sensors, computers, and thrusters to detect, track, and intercept enemy missiles. As a result, Brilliant Pebbles eliminated the SBI garage. Instead of several hundred large targets, Soviet ASATs would now have to contend with several thousand (about 4,000) small, light (45 kg; 100 lb), hard-to-find interceptors orbiting Earth in a constellation that would cover appropriate regions of the world.

Brilliant Pebbles also would lower the cost of implementing SDS Phase I. The SDIO would have the Brilliant Pebble interceptors mass-produced, so they would be relatively inexpensive (perhaps only $275,000 each). In addition, the Brilliant Pebbles interceptors would make it possible to eliminate one constellation of space-based sensors, further reducing the total cost. Brilliant Pebbles gained early support from Lt. Gen. Abe Abrahamson, head of SDIO, and Vice President Quayle. The decision to integrate Brilliant Pebbles into the SDS Phase I architecture came in early 1989, after Air Force Lt. Gen. George L. Monahan, Jr., succeeded Abrahamson as the SDIO director.[71]

Brilliant Pebbles created an excellent opportunity to sell the *SSX*. The small Pebbles meant that the SDIO no longer needed to place 100,000-pound (45,000 kg) laser battle stations in orbit. Rather, because each Pebble weighed about 100 pounds (45 kg), SDI launch needs shifted to a lighter, medium-lift rocket, such as that proposed in the *SSX* program.[72] Moreover, in order to set up Brilliant Pebbles, an unprecedented number of launches would have to take place. Because the *SSX* promised low-cost access to space, it could reduce the cost of implementing Brilliant Pebbles.[73] In general, in order for a reusable launcher to be cost-effective, it has to fly many times. Each flight has to amortize the vehicle's massive up-front capital expenses, such as research and development and manufacturing costs. "It's almost equivalent to buying a 747 or a C-5A for many hundreds of million dollars and using it five times a year. It is hellaciously expensive," Hoeser explained. "Each flight gets to be very, very expensive."[74] Brilliant Pebbles, then, with its need for hundreds of launches, was well suited for a reusable launcher.

It was Brilliant Pebbles, however, and not the *SSX*, that drew interest from the media, even as Graham, Hoeser, and the High Frontier organization publicized the *SSX* concept. A High Frontier presentation on Brilliant Pebbles and the *SSX* on Capitol Hill saw support for the "Star Wars" project divided along party lines, Hoeser observed, while "there was almost uniform support for the SSX concept." The response from the Defense Space Council, he noted, "ranged from dumbfounded to casual interest."[75]

A Home for the *SSX*?

Meanwhile, the National Space Council began its review of the *SSX* proposal. In May 1989, the council asked the air force to analyze the proposal, and the Space Systems Division of the Air Force Systems Command assigned the task to the Aerospace Corporation,[76] a civilian analysis arm of the air force. Although earlier Aerospace Corporation reports had dismissed the near-term feasibility of single-stage-to-orbit concepts, "this report was quite a bit different from ones previous," Gary Hudson explained.

> It concluded that the basic idea was feasible, with a few disagreements over technical approaches. In fact, the earlier version of the report [19 July 1989] was highly positive, while the final version [15 August 1989] was somewhat more restrained. It is instructive to note that a senior Aerospace executive authored two of the earlier reports that ridiculed both Phoenix and X-rocket. Since nothing had changed technically, the positive tone of the 19 July version was an embarrassment and the more muted endorsement of the 15 August version was necessary to obtain his release signature. . . In any case, the fact that Aerospace had essentially endorsed the concept paved the way for the initiation of the SDIO SSTO program.[77]

The Aerospace report's title, "Review of Pacific American Launch System SSX:Phoenix VTOL Concept," again treated Hunter's *SSX* and Hudson's *Phoenix* as the same vehicle. The report actually reviewed the feasibility of vertical-takeoff-and-landing single-stage-to-orbit transport in general, of which both the *SSX* and *Phoenix* represented specific designs. Because the Aerospace Corporation received more technical information on the *Phoenix* than the *SSX*, it became the reference vehicle in the Aerospace review. Hunter, however, believed that Jay Penn, the report's principle author, had confused the *Phoenix* and *SSX* vehicles. He wrote to Air Force Maj. Jess Sponable, then chief of the Space Applications Branch in the NASP Joint Program Office,

Wright-Patterson Air Force Base: "SSX is not Phoenix! To the best of my knowledge, everyone who knows anything about the situation agrees with this but Jay Penn. The Aerospace document is the only existent which refers to 'SSX:Phoenix.' Both Gary [Hudson] and I know there is no such thing."[78]

During July 1989, as the Aerospace Corporation was drafting its report and the *SSX* campaign was in full operation, the *SSX* concept began to draw the attention of certain individuals involved with the National Aero-Space Plane. Among them was Sponable. He wrote to the Aerospace Corporation with his personal views on the *SSX,* which he found to be generally feasible, although it "looked like a big tea pot."[79] Sponable had been associated with both Science Dawn and Have Region. Following declassification of Have Region in 1988, Sponable provided the Aerospace Corporation with a list of NASP and Have Region technologies applicable to the *SSX,* including slush hydrogen, composite materials, titanium alloys, and certain avionics subsystems.[80] As Daniel Graham observed in August 1989, "The SSX idea apparently has stirred up the dinosaurs." He saw what he thought were clear signs that the NASP program was interested: "The NASP contractors are now making noises about 'NASP-derived vehicles' which translates 'SSX.'"[81]

NASP-derived vehicle (NDV) was the term the National Aero-Space Plane program used to refer to the fully operational vehicles developed by industry from the fruits of NASP. NASP, after all, was an experimental program intended to lead to the design of a prototype, not an operational vehicle. Nonetheless, the NASP Joint Program Office had an Applications Directorate, which concerned itself with future applications of NASP technology as well as with analyzing and explaining the value of the NASP program to the public and Congress. As part of this work, the directorate also conducted extensive studies of NASP-derived vehicles.[82] Sponable, as part of the Applications Directorate, pushed single-stage-to-orbit vehicles from within the NASP program.[83]

The *SSX* was not intended to be a NASP-derived vehicle, though, and the *SSX* was not part of the NASP program. Jerry Pournelle strongly felt it should be kept apart from NASP. "I was talking to some NASP people Saturday at dinner," he wrote, "and they think SSX is wonderful, but of course it ought to be studied, and designed right, and done right—which really means the poor thing would have all kinds of 'requirements' glued onto it, would eventually get gold-plated, and sink into the Potomac mud."[84]

The *SSX* briefings and even Graham's contacts in key places, however, were not enough in and of themselves to convince people that a single-stage-to-orbit vehicle was feasible. As knowledge of the NASP program's problems spread, selling the single-stage-to-orbit concept became that much more difficult. Nonetheless, aiding and abetting the efforts of Graham, Hunter, Hoeser, the High Frontier, and the Citizens' Advisory Council were the results of two studies.

The first was the classified Have Region study. Its results persuaded several members of the National Space Council that a single-stage-to-orbit transport was now possible. It also excited a number of individuals in the SDIO, including Mike Griffin and Col. Gary Payton. Sponable briefed them on the Have Region study. Sponable explained: "The message they left with was: 'Hey, these aren't just nuts out of left field talking to us about SSX. There's a real government program, we've done real work, and it might be feasible to do this."[85] Looking back, Payton described Have Region as: "an eye-opening result to Griffin and me."[86] As a result, from Payton's perspective, the SDIO was ready to run with the project even without the endorsement of the Aerospace Corporation.[87]

The other study supporting the feasibility of rocket-powered single-stage-to-orbit transport was that carried out by the Aerospace Corporation. Graham and others obtained early drafts of the Aerospace report and showed them to Monahan, Griffin, Payton, and others in an effort to persuade them to support the *SSX*.[88] The report draft, dated July 19, 1989, gave a positive evaluation. The *SSX* "vehicle concept," the report stated, "has the potential to result in a reliable and economic method of space transportation that uses today's technology." Although the *SSX* would not be able to carry the payload weight claimed for it (16,500 versus 20,000 pounds, or 7,500 versus 9,100 kg), "our worst estimate of the uncertainties still shows useful orbital payloads."[89]

Buoyed by the report, Jerry Pournelle wrote to Quayle on July 24, 1989, pledging to have the *SSX* program on the schedule promised at the February 1989 briefing. The first flight test would occur in spring 1992, and the first orbital flight on Columbus Day 1992. The *SSX* program, Pournelle explained, needed a "firm commitment of $50 million before January 1990; it would be better to have $90 million in order to get subcontractors moving. No more than that is needed in 1990; in fact, more money might well be counterproductive, since the purpose of this project is to build space ships, not empires. We would prefer that the SSX program office be guaranteed to go out of business by 1993."[90] Little did Pournelle realize it at the time, but the *SSX* program office *would* go out of business, though not in 1993, and not because the program had accomplished its goals.

CHAPTER 7

The SDIO SSTO Program

The Aerospace Corporation's positive evaluation and the enthusiasm of SDIO officials Mike Griffin and Gary Payton paved the way for the *SSX* to become an SDIO project called the Single Stage To Orbit (SSTO) program. The SSTO program would fulfill the vision of a single-stage-to-orbit experimental program as outlined by Max Hunter, including its call for an emphasis on aircraft-like operations in a "faster, cheaper, smaller" program. The creation of the SSTO program and Phase I activities proceeded smoothly and quickly with a modest budget. The project faced no funding impediments or opposition. Meanwhile, however, the windmills of political change began to turn.

Getting with a Program

The *SSX* program got its "firm commitment of $50 million before January, 1990." Indeed, SDIO director George L. Monahan, Jr., indicated to Daniel Graham by mid-August 1989, shortly after the release of the final version of the Aerospace Corporation report, that he would commit SDIO money for the *SSX*, the funds to come out of the SDIO fiscal 1990 reserves.[1] This move gave the project stealthy underwriting that would be invisible to budgetary radars on Capitol Hill. Monahan later remained a solid supporter of the SSTO program. He wrote to Graham in January 1990: "I remain firmly committed to exploring single-stage-to-orbit (SSTO), rocket-powered vehicles in 1990. Unfortunately, we will not be able to invest as much as we had originally planned, but I am still eager to get the effort underway. Our work will focus on an experimental, technology demonstrator, with initial flights slightly more than four years from now."[2] Although SDIO originally had planned to make awards to more than one contractor in Phase II, SSTO Program Manager Lt. Col. Hayward P. "Pat" Ladner told a reporter that the agency "probably" would select "just one."[3] The money simply was not there.

But those budgetary difficulties were in the future. In 1990, as Mike Griffin later recalled: "He [Monahan] asked me and my deputy, Gary Payton, to make the program happen. We, of course, were not really surprised, because we had been pushing for it from the other end, from underneath. The program didn't walk into our door as a surprise request from the general. Nonetheless, he was the one who had to make the budgetary place for it, and he did that."[4]

The program had not been a surprise to Griffin and Payton, or Monahan, because of the briefings conducted by Graham, Hunter, Hoeser, and High Frontier. Monahan was the target of several letters from Graham, who knew Monahan on a first-name basis, as well as a twenty-minute briefing by Graham on July 14, 1989.[5] At the same time, Graham, Hunter, and Hoeser briefed Mike Griffin, SDIO deputy for technology, and his assistant, Gary Payton, in July and August 1989.[6] Griffin was one of two SDIO deputies who reported directly to the director. The deputy for systems was in charge of architectures and battle space management; the deputy for technology, Griffin, oversaw all research, development, and flight tests. Griffin had taken over as SDIO deputy for technology in January of 1989 after already having conducted SDI work while at the Johns Hopkins Applied Physics Laboratory. Among the larger SDIO projects he had overseen was the Advanced Launch System, an effort to develop a new Saturn V–class booster to launch large space-based lasers for the Strategic Defense Initiative, or to travel back to the Moon and onward to Mars.[7] Payton, a former air force payload specialist on the shuttle orbiter *Discovery,* started at the SDIO in 1986. Before working under Griffin in 1989 on the ALS and other projects, he worked at the SDIO on sophisticated infrared sensors that would form part of a space-based defense network.[8]

By January 1990, planning for the SDIO single-stage-to-orbit program was well under way. Monahan wrote Graham: "Current plans include releasing an announcement in the *Commerce Business Daily* during February, followed by a bidders conference here in Washington, then an RFP [Request For Proposals], all aiming at a contract award in May or June 1990."[9] The *Commerce Business Daily* announcement did not appear until March 26, 1990. Nonetheless, with that announcement, the SDIO's Single Stage To Orbit program was officially underway.

The *Commerce Business Daily* announcement described the program as the SDIO then conceived it. Phase I would consist of "research into single stage to orbit [vehicles]," specifically, "design trades, identification of the critical path technology, and a risk reduction demonstration along that critical path." SDIO expected to make not one but "multiple contract awards of up to four contracts." Phase I would last only ten months. This was a "fast" program. SDIO then would hold a competition for Phase II, the manufacture and flight testing of a suborbital vehicle. The SDIO would select "the most promising concept(s)" for Phase II, the idea being to select one or two contractor designs.[10] The *Commerce Business Daily* announcement did not mention that the SDIO planned a Phase III for the design, development, construction, and flight of a full-scale orbital prototype.

A fuller description of the program appeared in the Request for Proposals (RFP), SDIO84-90-R-0007, issued on May 9, 1990. The SDIO contract officer, Lt. Daniel L. Downs, held a workshop for potential industry bidders at Bolling Air Force Base on May 16, 1990. Assisting him were Mike Griffin and Pat Ladner, Director of Special Operations. Ladner had replaced Gary Payton, who was attending the Air War College, as SSTO program head. Before leaving, Payton had helped to write the RFP's Statement of Work.

"Gary wrote, and I edited, then he rewrote, and I re-edited, and between us we put together a statement of work of a couple of pages setting out what we wanted the DC-X [that is, the Phase II vehicle] to demonstrate,"[11] Griffin explained. In drafting the Statement of Work, Payton bore in mind advice given him by astronaut T. K. Mattingly, whom he had met as the commander of Payton's shuttle flight.[12] "One of the things T. K. said was: 'Get the government out of the business of specifying the details. Just lay out your mission requirement, and let industry figure out how to design the bird, how to achieve the mission requirement.' Great idea! It was totally contrary to the practices of the previous 30 or 40 years of history. I said, 'Yeah, that's a great idea. I mean, industry has been crying for decades that government over-specifies everything. Let them have complete design authority. Boom. Good idea. We did that."[13] Griffin also emphasized that the parameters for the SDIO's single-stage-to-orbit vehicle were left "very wide open, quite deliberately."[14] This philosophy made the SDIO SSTO program even more different from previous aerospace programs.

The Phase I Statement of Work spelled out the SSTO program's stages. Phase I would see multiple research awards to industry. Phase II would consist of the manufacture and flight-testing of a suborbital technology demonstrator (manned or unmanned, at the contractor's discretion). Phase III would involve building and flying a manned full-scale experimental prototype capable of achieving orbit, with the first flight scheduled to take place in fiscal 1994 (two years later than what Pournelle had told Quayle). According to the Statement of Work, the Phase II vehicle "will be flight tested to perform experiments focusing on critical technologies," while Phase III vehicle flights' "emphasis will be on demonstrating reduced manpower requirements for readying the vehicle for flight."[15]

Although one could read the Statement of Work to mean that the program's focus would shift from technology development in Phase II to operational goals in Phase III, the primary program objective was an operational, manned, reusable single-stage-to-orbit vehicle "with aircraft style operations." From beginning to end, as Max Hunter and his supporters within the Citizens' Advisory Council advocated, the emphasis was to be on operations, not technology.

"What we were after primarily," Griffin explained,

> because we had a limited amount of money, was a high degree of operational reuse. . . . Gary's and my primary concern was to try to escape the trap that you can see if you're on the inside of an Atlas, Delta, Titan, or Shuttle launch: the hundreds and thousands of people. Look at the airplane. They manage to dispatch airplanes, even very sophisticated airplanes, with many fewer people than it takes to dispatch a launch vehicle. You look hard at that, and you see that the airplane doesn't look any simpler than the launch vehicle. It's somewhat different, but it doesn't look a lot simpler. So why does it take so many people for rockets, and it doesn't for airplanes? When we landed guys on the Moon, it was in a pretty complicated vehicle. It was as simple a vehicle as they could make, but it was still pretty complicated. Somehow or other, though, two guys managed to power up, check out, launch, and fly the spacecraft from the surface of the Moon with no ground support at all. Well, they did it because that was how they had to do it. . . . So it leads you to ask the question: If they didn't have any ground support people at Cape Canaveral or at Vandenberg Air Force Base, would they find a way to launch rockets without them? Of course, they would. . . . Gary and I were convinced that, if we had only a little bit of money, what we wanted to demonstrate was this sort of operational turnaround and reusability.[16]

The SDIO's thinking was rather simple. Have Region had just spent over $50 million on a technology study, and NASP was spending $2.5 billion mostly on structures and materials. Adding the relatively small amount available for the SDIO SSTO program to further technology development would have been a waste. The Statement of Work reflected this thinking: "This program is intended to capitalize on recent and near-term advances in launch vehicle propulsion, structures, and operations deriving from the NASP, ALS, and SDIO materials and structures programs. The restriction to near-term is intentional."[17] Instead of technology, the program would address the key issue of the "standing army" needed to operate and maintain launch systems, such as the thousands needed for NASA's shuttle, which inflated launch costs.[18] Consequently, the Phase I Statement of Work stipulated that: "Design shall focus on operational simplicity and minimizing vehicle processing requirements. The highest priority shall be that of meeting the vehicle turnaround goal."[19]

The RFP divided the ten-month-long Phase I into two parts. During the first four months, each contractor would explore all of the basic single-stage-to-orbit architectures: vertical takeoff and landing, horizontal takeoff and landing, and vertical takeoff and horizontal landing. The RFP, and specifically the Statement of Work, did not suggest that the SDIO would give preference to any one approach. Nonetheless, Gary Hudson felt that the SDIO had a clear preference for vertical take-off-and-landing (VTOL) vehicles. "While it was widely understood in the community that the program managers at SDIO favored the VTOL concepts typified by the SSX," Hudson wrote, "pressure from at least one of the potential contractors forced SDIO to open the competition to all types of SSTOs [single-stage-to-orbit vehicles]."[20] Major Sponable, an active advocate of single-stage-to-orbit vehicles in the NASP Applications Directorate, argued for keeping the SSTO program competition open to all vehicle concepts.[21]

During the following six months of Phase I, each contractor would select a preferred architecture and refine their design. The design had to be "essentially" capable of single stage to orbit, meaning "takeoff or liftoff assistance is acceptable as long as the above design goals are attained." By including this clause, the Statement of Work opened the door for Boeing's *RASV*. The design goals stipulated in the Phase I Statement of Work emphasized operations, not technology development. The vehicle should have "fail-safe operation, engine-out capability during all portions of powered flight, and flight crew escape capability during ascent and entry." Moreover, the vehicle should be able to handle a "launch rate surge to double the routine launch rate, to be maintained for 30 days."[22] These were operational, not technological goals.

At the end of Phase I, according to the Statement of Work, the SDIO would review contractor study results and designs, draw up a Statement of Work for Phase II based on those results, and open competitive bidding for a Phase II contract. To help the SDIO evaluation team to assess the capabilities of the contractor vehicles, they created a "design exercise" against which to compare them. It assumed a craft capable of lifting 10,000 pounds (about 4,500 kg) of payload weight into a circular orbit at 400 kilometers, at a 90° inclination, and which launched from the continental United States. Payton wanted the design exercise to be included in the bid, but the final product did not have to be a vehicle that accomplished the design exercise goals.[23]

The design exercise parameters were not random but related specifically to Brilliant Pebbles. Moreover, a classified appendix to the Statement of Work provided details on the Brilliant Pebbles payloads that the full-scale launcher would loft into space. Linking the SSTO program to Brilliant Pebbles was the idea of not just Daniel Graham, but Gary Payton as well.[24] Thus, just as the campaign to sell the *SSX* had linked it to the SDI and Brilliant Pebbles, so the SDIO coupled the SSTO program to Brilliant Pebbles and, by implication, to cold war preparedness. Although all of the winning contractors included single-stage-to-orbit designs that launched and landed vertically like the *SSX*, each differed from the *SSX* in the details of its basic design. The influence of Max Hunter and the *SSX* came through strongest in the SSTO program's emphasis on aircraft-like operations. Although the *SSX* no longer played a part in the development of the SSTO program once Phase I started, Hunter, under the business name of SpaceGuild, eventually participated in the McDonnell Douglas team that won the Phase II contract.

Phase I Concepts

The Phase I Statement of Work stipulated that proposals were due July 2, 1990, "so industry would not force their proposal teams to work over the July 4 holiday," Payton explained.[25] Evaluations would take place July 9 through August 9, 1990, and the award of the Phase I contract(s) would take place on August 15, 1990.[26] SDIO received six proposals and made four awards ranging from $2.4 to $3 million on August 15, 1990. The winners were Boeing, General Dynamics, McDonnell Douglas, and Rockwell International. The two unlucky proposers were Grumman and Third Millennium, Inc., both of which favored vehicles that launched and landed horizontally.[27] The four contractors undertook the mandated studies of various single-stage-to-orbit configurations, selected their preferred vehicle configuration, and presented their results in December 1990. Both McDonnell Douglas and General Dynamics settled on vertical-takeoff-and-landing configurations. Boeing proposed an improved version of its horizontal-takeoff-and-landing sled-launched *RASV*, while Rockwell selected a vertical-takeoff-and-horizontal-landing vehicle.[28]

Next, the SDIO subjected these design concepts to review by outside agencies, the Aerospace Corporation and NASA's Langley Research Center. The Aerospace study began in January 1991 under the direction of Jay Penn. Their study provided general observations on, and even praise for, the SSTO program, specific comments on the contractors' approaches to Phase II planning, and a list of tasks that the Aerospace Corporation could carry out for the program.[29] "I think," Penn wrote, "the SSTO Program is potentially one of the most interesting and revolutionary projects this nation has opted to pursue." Moreover, he was "convinced, based on contractor and independent Aerospace [Corporation] studies, that SSTO is feasible and that operability and reliability goals can be met and that the cost and capability benefits are well worth the development risk."[30]

Rather than endorse a single contractor design, the Aerospace review proposed that a consortium carry out Phase II. The SDIO would approve two vehicle design contractors through Preliminary Design Review (PDR), then would pick a single winner. The losing contractor would become a subcontractor to the winner. Each contractor would contribute its best features. General Dynamics, for example, seemed to have the best flight operations and technologies, while Boeing offered strength in ground processing and checkout based on commercial aircraft activities. "The consortium approach," Aerospace argued, "would enable the government to have maximum utility of each of the contractors' strengths and would also provide business opportunities for all of the contractors. This consortium option should be considered."[31] The same consortium arrangement was under study by the NASP program. The SDIO, however, did not want another large-scale cold war program.

The Aerospace review also ranked the contractors' designs. McDonnell Douglas's appeared "strongest overall." The review declared that the vertical-takeoff-and-landing design was "a winner" and had the "best overall risk reduction program," as well as a "strong focus on operations." General Dynamics, in turn, had the "best treatment of operations," but its overall vehicle design was "poor." Rockwell had a "strong design option and good operations emphasis," but its design had an "uncomfortably low design margin" and assumed an engine performance that was too high. Boeing's design focused on structures and engine upgrades that would be easily attainable and had an "excellent treatment of commercial aircraft processing technologies." However, it appeared to be "least likely to achieve SSTO operability objectives."[32]

NASA's Langley Research Center conducted a separate review in response to Mike Griffin's request on February 22, 1991.[33] William Piland, director of the Space Systems Division, and Alan Wilhite, head of the Systems Analysis Branch, were responsible for the review.[34] The major Langley contributors were Richard W. Powell, study director, and Douglas O. Stanley, lead engineer, both with the Vehicle Analysis Branch.[35] The Langley team considered only three of the four contractor designs. (Although the Boeing Defense and Space Group made a Phase I final report in June 1991,[36] they elected not to bid. The Aerospace Corporation review had been the coup de grace for the *RASV*.)

SSTO Program Manager Pat Ladner "requested that the LaRC [Langley] team concentrate on deficiencies of the concepts and design methodologies. Thus, many of the detailed comments in the present analysis are directed towards the shortcomings of the vehicles," the report explained. It was not intended to be a comprehensive evaluation of each vehicle design. "Only those issues and disciplines were addressed for which LARC had sufficient time and experience to examine." Moreover, "the time limitation [about two months] was further complicated by the constantly changing contractor configurations, different design and technology assumptions, and different selected design reference missions."[37]

The Langley study stated up front: "This report neither compares nor recommends any particular configuration for follow-on SDIO Phase II contracts." It was a stance of objectivity, rather than partisanship, and the evaluations contained in the report were professional opinions. As Doug Stanley explained: "We purposely did not say: 'Pick this contractor or this contractor.' We just showed the strengths and weaknesses of each, and we let the chips fall where they may."[38] The Langley conclusions were more pessimistic than those of the Aerospace Corporation: "The overall conclusion of this LARC study is that none of the current contractor concepts have matured sufficiently to be a viable concept." Each concept had serious unresolved flight control, thermal protection system, and cryogenic tank technology issues.

The Langley team was not against single-stage-to-orbit vehicles per se. Rather, they concluded, although "none of the contractors demonstrated concept viability, it is the opinion of the study team that a single-stage-to-orbit vehicle can be developed for small payloads using a fail-safe design philosophy." Their study, like that of the Aerospace Corporation, pointed out the need for further development of the aerospike engine. They also highlighted other technologies that required maturation for a single-stage-to-orbit vehicle, such as reusable, light-weight cryogenic tanks, durable thermal protection systems, and automated, accurate flight control, guidance, and navigational systems.

The three vehicle designs that Langley studied varied in terms of mission capability and general design. The General Dynamics Space Systems Division proposed a vertical-takeoff-and-landing vehicle called the Model 107. It was a 15-foot (4.5-m) cone powered by a Rocketdyne aerospike engine and was capable of delivering a 15,800-pound (7,200-kg), six-person crew to space station *Freedom* from the Kennedy Space Center.[39] The *Delta Clipper* (versions *DC-X* and *DC-Y*), proposed by the McDonnell Douglas Space Systems Division, was a cone-shaped, vertical-takeoff-and-landing vehicle. The prototype *DC-Y* would be able to carry a 6,700-pound (3,000-kg) payload and two crew members to a 90° inclination, 216 nautical-mile (400-km) circular orbit from Vandenberg Air Force Base, or a 17,300-pound (7,900-kg) payload and two crew members to Space Station *Freedom.* The Langley study found that McDonnell Douglas used "a number of subsystem technologies that contribute to reduced maintenance procedures." Also, the contractor's concept met the SSTO program's operational turnaround and surge goals by a large margin.

The Rockwell International concept was a winged, rocket-powered vehicle that took off vertically and landed horizontally on conventional landing gear like the NASA shuttle. It used the rocket engine only for ascent and would glide in the event of a major propulsion system failure. The Rockwell design would deliver 17,200 pounds (7,800 kg) of payload and two crew members to space station *Freedom.* Among other problems the Langley team found fault with was the Rockwell thermal protection system: "The adhesively bonded blanket thermal protection system (TPS) does not appear to be fail-safe or all-weather. Additional technology development will be required to demonstrate the feasibility of the blanket TPS use on the windward surface. The waterproofing required after each flight will also complicate operational procedures."[40]

Phase II Statement of Work

With the completion of the Langley study in the first half of 1991 and the contractors ' presentations of their results in June 1991, the SDIO opened the Phase II competition on June 5, 1991, with Request for Proposals

SDIO84-91-R-0008. The cover letter, addressed to the "Prospective Offeror," announced that: "The Government anticipates the award of up to two contracts for the Phase II technology demonstration. . . . It is the Government's intention to award up to two contracts within the $52M budgeted for Phase II to the best overall offeror(s) on the basis of the evaluation criteria set forth in Section M." The decision had not yet been made to award a single contract. The cover letter also emphasized the SDIO's hands-off approach to the project: "Contractors are encouraged to be innovative and creative with minimal Government guidance and direction. That type of atmosphere, from both the Government and the selected contractor(s) is the goal of this SSTO acquisition."[41]

The Phase II Statement of Work neither dictated nor described the vehicle to be designed, built, and test flown during Phase II. Rather, it described the fully operational vehicle that would be built after Phase III. Although Boeing was not participating, the Statement of Work left the door open for the *RASV* or a similar design employing "some forms of takeoff enhancement . . . as long as inclusion does not seriously compromise turnaround, orbit access, or other major characteristics of the system."[42] This was one of the few hardware requirements laid out in the Request for Proposals. For the most part, the short, thirteen-page Statement of Work defined the full-scale vehicle in terms of operational performance, not technology.

That vehicle would be capable of both manned and unmanned operation to and from all the usual orbits. In the absence of a human operator, it would be capable of performing normal programmed operations. It also would take off and land "without damage" in inclement weather conditions and in a crosswind of 25 knots with gusts to 35 knots, would be capable of remote "pilot" ("virtual cockpit") control, and would be capable of operating despite a number of system failures. Max Hunter's insistence on flight-abort capability rang through: "Intact abort is desired in all noncatastrophic failure cases. . . . The vehicle must be able to accommodate complete loss of thrust from one or more engines during powered flight such that the probability of catastrophic vehicle loss is minimized."[43]

The program's main goal was set forth in the Statement of Work. "By balancing design, operational and maintenance factors, the SSTO will drive system costs to their lowest possible level." In order to achieve low-cost operation, the vehicle had to require minimal maintenance between flights and had to allow rapid turnaround "by today's standards." "The [Phase II] vehicle must be capable of being turned around," the Statement of Work stipulated, "within 7 days of landing with the expenditure of no more than 350 man-days." The Phase III prototype could require more than 350 man-days. For the full-scale vehicle, the amount of time a mission had to spend between initial request and launch should be less than 30 days. In addition, the vehicle had to have "launch rate surge capability" (doubling the routine launch rate) "for a minimum of 30 days."[44]

The Statement of Work also stipulated how much payload weight the operational vehicle would be expected to place in orbit. The goal was "payloads as large as 10,000 lb (4,500 kg) for polar launch azimuth into 100-nmi [nautical mile] circular orbit." The desired payload bay would be "as near as practical to 15 ft (4.5 m) diameter by 30 ft (9 m) length. [However,] This is not a hard requirement—merely an indication of desirable range and potential future missions."[45]

The altered orbital specifications for the full-scale, single-stage-to-orbit craft were no longer tied to Brilliant Pebbles specifically or the Strategic Defense Initiative in general. In the year between the Phase I (May 1990) and Phase II (June 1991) requests for proposals, the world had changed, and so had the need for the SDIO's SSTO program. The Phase II Statement of Work specified that the full-size vehicle would "perform Space Transportation System (STS)-like [space shuttle-like] operations for rendezvous, docking, and also propellant refueling to either active (cooperative) or passive (noncooperative) spacecraft." Other missions included deployment of satellites and other payloads in low and geosynchronous orbit; space station and space shuttle support; satellite recovery, replacement, and improvement; and the Space Exploration Initiative (SEI), President George H. Bush's grandiose plan to return to the Moon and to land astronauts on Mars. As examples of satellite constellations the full-scale single-stage-to-orbit vehicle would service, Program Manager Pat Ladner cited the air force's Navstar Global Positioning System and Motorola's Iridium, a set of satellites expected to provide global mobile telephone service.[46]

The full-scale vehicle, though, would not be built until after the completion of Phases II and III. During Phase II, the contractor(s) would complete design of a scale suborbital vehicle and all support infrastructures (such as the launch pad); demonstrate critical hardware, software, and operational concepts; and begin plans to manufacture the Phase III prototype vehicle. The Phase II craft would be an "X" vehicle, the Phase III craft a "Y" vehicle. The "Y" indicated a prototype. The Statement of Work laid out a general schedule for completion of both Phases II and III. The SDIO would select the Phase II contractor(s) in August 1991. The Preliminary Design Review (PDR), that is, when the SDIO would approve the vehicle design, would take place six months later (February 1992). The Critical Design Review (CDR) would take place in May 1993, with "probably bimonthly" progress and status reviews between the PDR and the CDR. Phase III would be the construction and flight testing of a prototype. The first Y-vehicle suborbital test flight would take place in 1995, and the first orbital test flight in 1997.[47]

In order to determine which contractor or contractors would receive Phase II funding, the SDIO attended presentations in May and June 1991 by the three competing contractors: General Dynamics, McDonnell Douglas, and Rockwell International.[48] The SDIO then declared McDonnell Douglas the winner. The Langley study had given McDonnell Douglas high marks on operations (the key to the SSTO program) and highest all-around marks. The Aerospace Corporation review also acknowledged them as having the best overall design. According to Mike Griffin, who was responsible for selecting the winner, the SDIO reached its decision based on the firm's ability to build a full-scale single-stage-to-orbit vehicle and other criteria. "The contractor had to demonstrate," he explained, "how his basic vehicle could and would scale up to the next generation of vehicle. We weren't just interested in the DC-X level. That was something we viewed as a start, but it had to be able to grow. We thought that the General Dynamics proposal, from the point of view of reentry and heat transfer, didn't grow well, if I recall. Rockwell management at that time didn't appear to be fully in support of the project. . . . McDonnell Douglas's management clearly was behind it."[49]

Jess Sponable, who chaired the technical evaluation board, favored the McDonnell Douglas proposal because they "were the only ones that really were going to fly something and fly it again and again. An initial proposal by G.D. [General Dynamics] was to actually build one, but not fly it. They later proposed they would fly it one time. Even though it would've gone faster than the McDonnell [Douglas] vehicle, we couldn't quite justify building this thing and flying it just one time." Above all, Sponable and the SDIO were looking for a flight test program. "What we needed to do was demonstrate how you integrate a set of technologies into a flight vehicle and do the aircraft-like ops [operations]. The next step would be to take the advanced technologies and build a flight-weight vehicle."[50]

As for the selection of McDonnell Douglas and the *Delta Clipper,* Sponable explained: "We didn't think any of the others would survive in the bureaucracy. We felt we needed a demonstration of aircraft-like operability, the ability to really fly this thing multiple times and to operate it like an aircraft." The other option the evaluation board seriously considered was the aerospike engine demonstration program proposed by Rockwell. However, the SDIO was not interested in investing in a technology development program that had no application, and the technical risks were too high.[51]

On August 16, 1991, the SDIO awarded the Phase II SSTO program contract to McDonnell Douglas. It was a $58,904,586 contract to design, build, and flight-test the one-third-scale, suborbital, *Delta Clipper Experimental (DC-X)* reusable rocket within twenty-four months (by August 1993).[52] The name honored the firm's successful *Thor/Delta* rocket and recalled the famous nineteenth-century commercial clipper ships. The McDonnell Douglas SSTO program team saw the *Delta Clipper* as opening the "space trade routes in the same way that the Yankee Clipper ships opened the sea trade routes." The needs of the commercial launch industry thus were integral to the thinking of the McDonnell Douglas *DC-X* team.

The Weasel Works

McDonnell Douglas senior management initially had endorsed the firm's bid for the SSTO program only lukewarmly, but once money appeared available to build the Phase II vehicle, they provided "good support" to put together a proposal.[53] In order to undertake the daunting task of designing and building the *Delta Clipper* in only twenty-four months, the company formed the Rapid Prototyping Department. The term "rapid prototyping" came into vogue during the 1990s to identify new computer-aided design and manufacturing (CAD/CAM) methods.[54] Beginning in 1995, "rapid prototyping" spawned a pair of journals, the *Journal of Rapid Prototyping Technology Abstracts,* and the *Rapid Prototyping Journal,* as well as numerous international conferences sponsored by such diverse groups as France's Centre National de la Recherche Scientifique (CNRS), the International Society for Optical Engineering (SPIE), the European Optical Society (EOS), the Directorate General for Science, Research, and Development of the Commission of the European Communities, the Society of Manufacturing Engineers (and its Rapid Prototyping Association), and the Computer Society of the Institute of Electrical and Electronics Engineers (IEEE).[55]

McDonnell Douglas, however, used the term to describe a certain management approach to manufacturing that integrated various computer-assisted activities as well as a number of other stratagems that included collocation of engineering, design, and manufacturing; small management teams; and reduced paperwork. The Rapid Prototyping Department merged "integrated product team development" from Total Quality Management (TQM)—which corporate McDonnell Douglas endorsed at the time—with characteristics of the "fast-track" approach, such as delegated decision-making, short management lines, and collocated team members.[56] The firm's "rapid prototyping" thus combined the essentials of a "fast-track" approach with TQM. In this sense, "rapid prototyping" signified a more general management system than the more limited, but more generally accepted, definition of "rapid prototyping."

The term "fast track" is greatly overused. One can understand McDonnell Douglas's "rapid prototyping" as an industrial version of the "faster, smaller, cheaper" style of project management practiced by the SDIO. "Fast-track" approaches were not new to the firm. The X-Shot, for example, dated back many years. Program Manager Marvin Mark in the St. Louis division used the X-Shot during a competition with Boeing for the YC-14 and YC-15. "We built two airplanes," Pete Conrad, McDonnell Douglas manager and former astronaut, explained. "We just sort of drew a picture, and hammered metal, and drew some more pictures, and hammered some more metal. That was the X-Shot way of doing it, rather than trying to go through a full-blown program."[57] Another appropriate term for McDonnell Douglas's "rapid prototyping," and one commonly used in the space and aeronautics industries, is "skunk works," a generic term for shops managed in this fashion. The name derives from the popular name for Lockheed's Advanced Development Project.

In an ongoing effort to capitalize on "rapid" or "fast" management styles, aerospace firms often created so-called advanced development offices. Boeing, for example, had and still has its Phantom Works. Because their existence and activities are usually classified, little is known about them. Lockheed's Advanced Development Project has received the widest attention, particularly because of its self-promotion. The original "Skonk Works" appeared in Al Capp's "Li'l Abner" comic strip as a place where Appalachian natives combined skunks, old shoes, and other bizarre ingredients to concoct funky brews suited to a variety of purposes. The Lockheed "Skunk Works" has concocted such formerly secret aircraft as the U-2; the F-80 Shooting Star, the first aircraft to win an all-jet battle; the F-104 StarFighter; the SR-71 Blackbird; and the F-117A Stealth Fighter, famed for its performance in the 1991 Persian Gulf War.[58]

The pioneering effort of the Skunk Works in developing this management approach has received wide acceptance, but it is not necessarily accurate, because the nation's major aircraft firms all have had special organizations at one time or another. Nonetheless, aerospace companies using a "rapid" or "fast-track" management style often call their shops skunk works, giving the term a more generic meaning. Thus, for better or worse, the Lockheed Advanced Development Project has become the quintessential "fast-track" shop. As such, its internal operating rules, called Kelly's Rules[59] after Skunk Works founder, Clarence L. "Kelly" Johnson, have come to define the "fast-track" management approach (see appendix).

The name "Rapid Prototyping Department" struck Max Hunter as lacking a certain "pizzazz." He proposed instead the Weasel Works. "Weasels," Hunter explained, "are lean, mean fast little animals. . . . Weasels are exceptionally strong for their size. . . . Weasels can change colors to Ermine white as appropriate (in winter). . . . Weasels are clever—hence the common belief you can 'weasel around' anything. (sounds like our team!)"[60] Despite its novelty, and alliteration, the name did not catch on. In fact, McDonnell Douglas never considered using the name.[61]

Although today everyone "knows" what a "skunk works" or a "fast-track" or a "rapid" management approach is, there is little agreement on what those terms actually mean. A common misconception is that large-scale projects carried out to meet urgent deadlines, such as the Manhattan Project that developed and built an atomic bomb during World War II, are examples of "fast-track" management. The management of the vast Manhattan Project, with operations spread across the country and similar high-priority, large-scale programs, had no more to do with "skunk works" shop operations than did NASA's Apollo program to put an American on the Moon by the end of the 1960s.

Nonetheless, these large-scale projects did share one characteristic with Lockheed's Skunk Works and many other urgent programs: an abundance of government underwriting. Often, in order to expedite a program, government agencies have funded special programs lavishly to resolve technical bottlenecks and to meet program deadlines. Examples abound. Max Hunter, for instance, while working for Douglas, built the *Thor* missile in only nine months. The project was not "aimed at low cost, specifically," meaning substantial government underwriting facilitated its completion. Moreover, Douglas achieved this remarkable success without using the Skunk Works as a model.[62]

Unlike the Thor, Manhattan Project, and Apollo programs, the SDIO's SSTO program required that McDonnell Douglas build its *DC-X* vehicle in twenty-four months without an "open checkbook." The project, after all, was not a national priority, like a fighter aircraft or spy plane. Decisionmaking was delegated, management lines were short, and team members were collocated. Implementing a "fast-track" approach would help the firm achieve that daunting deadline within budget. McDonnell Douglas's "rapid prototyping" also differed from the usual "fast-track" project in that it borrowed from Total Quality Management.

The Rapid Prototyping Department used "integrated product team development" that had been developed for commercial industry, not the classified world of the cold war. "Integrated product team development" integrated all of a project's functions—such as analysis, design, procurement, testing, management, manufacturing, and cost control—into a "team" linked together by teamwide communications and a "commitment" to the project.[63] "Integrated product team development" came from a movement in management theory called Total Quality Management (TQM) and associated with its founder, W. Edwards Deming.[64] Deming's management ideas, concisely contained in his "fourteen points for management,"[65] failed to take root in his native country, the United States, but found a home in Japan. Japanese industry was still recovering from World War II, and the country had a wide reputation as a manufacturer of shoddy goods throughout the 1950s. Following adoption and implementation of Deming's fourteen points, international regard for Japanese manufactured goods changed. "Made in Japan" was no longer a name for poor-quality merchandise, as Japanese consumer products, especially consumer electronics, infiltrated and dominated one market after another.

Once Japanese industry experienced a turnaround assisted by the adoption of Deming's fourteen points, United States exports, which had enjoyed some success in foreign markets, began to suffer, starting around 1968. American industry began to ask itself whether or not it should consider adopting what it perceived to be Japanese management techniques. By the 1980s, the United States imported more goods than it exported, becoming more dependent than ever on foreign-made products. American industry then turned to Deming's fourteen points and a host of management gurus who preached the gospel of "excellence," "quality," "Quality Circles," and "teams," based on Deming's work.

Deming's fourteen points and TQM, which is based on it, stressed the importance of manufacturing quality goods and of providing quality services. High-quality products and services would win a larger market share and would create more profits and more jobs. Deming and his TQM disciples made satisfying the customer the focus of management attention, but moved the locus of where this occurs from its traditional place, in sales and marketing, to the manufacturing process. Deming's emphasis on managing the manufacturing process put his ideas in the same conceptual tradition as scientific management, also known as Taylorism after its founder, Frederick Winslow Taylor (1856-1915). Scientific management also focused on the actual manufacturing process, on both labor and machinery, in order to increase efficiency and maximize profits. Both also measured human and mechanical activity and performed quantitative analysis of labor and production machinery. However, while scientific management introduced the efficiency expert, Total Quality Management brought industry the Quality Circle.[66]

In any case, both scientific management and TQM theories brought the attention of management to the production line. They were reminders (albeit unknowingly) that manufacturing is done in a shop culture. This is in contrast to most management theory, whose focus is on organizing and administering large organizations, namely, the modern corporation. Businesses have not ceased to grow in size and complexity, and professional managers have tried to achieve the forbidding task of maintaining order and sustaining growth simultaneously. As a consequence of attempting to deal with this ongoing expansion and increasing complexity, corporate organizations have acquired thicker and more convoluted hierarchical layers, and both paperwork and regulations have multiplied.[67] A parallel growth in managerial and regulatory complexity also took place within the federal bureaucracy and was the target of President Reagan's harangue that government had "overspent, overestimated, and overregulated." It is within this context of bureaucratic size and complexity that the desirability of a "fast-track" approach becomes apparent.

Shops operating in a "fast-track" or "skunk works" mode are not new. They are, in fact, rather old, probably as old as shop practice itself. Indeed, the well-documented operation of Thomas Edison's West Orange facility during the 1880s and later resembled that of a modern "fast-track" shop. Engineering, design, and manufacture all took place in the same building, with the thinnest levels of management and almost no paperwork.[68] Within the aerospace industry, one finds early examples of collocating engineering and manufacturing. The original Douglas Space Division plant in Santa Monica, according to veteran employee Bill Gaubatz, had engineering and manufacturing operating side by side: "Our offices were right above the assembly line. Speaking of your construction work in the background: you heard the riveters going from morning to night."[69] By seeking (unconsciously) to recreate the manufacturing and engineering environment typified by Edison's West Orange facility, or even the early aviation industry, American firms have attempted to defeat the complicating levels of management and paperwork that are part and parcel of the rise of large corporate organizations both inside and outside the military-industrial complex. That was certainly the objective of the McDonnell Douglas SSTO Rapid Prototyping Department.

The Rapid Prototyping Department came together in September 1991, a month after the firm won the Phase II contract, under the direction of Bill Gaubatz, program manager, and Paul Klevatt, deputy program manager. Klevatt set up the department after heading a team that had brainstormed ways to facilitate key processes. They concentrated on both computer software and electronic hardware and tried to maintain a high product quality while accomplishing all program milestones. Klevatt also had to convince management that these processes actually could be improved. While Klevatt ran the manufacturing and flight test operations, Gaubatz focused on keeping the program funded and dealing with other external concerns. McDonnell Douglas also planned to use the Rapid Prototyping Department for the subsequent SSTO program Phase III vehicle, the *DC-Y.* The DC-1 would be the fully operational model of the *Delta Clipper,* expected to be ready for fleet operation by 1998.[70]

The SSTO Rapid Prototyping Department reported directly to the vice president for Advanced Product Development and Technology (APD&T). It operated according to a number of simple rules that were typical of a "skunk works" shop. For example, key design decisions would take no more than five working days, though the goal was "always one day or less." Computers executed engineering drawings and specifications, as well as hardware procurement, and department personnel would streamline the procurement of "catalog-type" parts and services. To meet the early delivery schedule for the *DC-X,*

Fig. 3. Paul Klevatt, McDonnell Douglas deputy program manager and organizer of the Rapid Prototyping Department. (Photograph courtesy of Paul Klevatt.)

concurrent engineering became a requirement, not an idealistic design philosophy. Aerojet, a *DC-X* subcontractor, also used a rapid prototyping approach and small concurrent engineering teams. Both Aerojet and McDonnell Douglas electronically shared a common database of files. Employees, equipment, and decisionmaking were collocated within the same building. The vehicle would be built and integrated by about a hundred people. Klevatt and his team eliminated or reduced paperwork, though without the advantages of today's more sophisticated electronic tools or software aids. Nonetheless, by the time the *DC-X* was built, they had reduced by 70 or 80 percent the amount of time it took to write software code.[71]

The *DC-X*

The *DC-X* built by McDonnell Douglas's Rapid Prototyping Department was an evolved version of the original concept. As described to the SDIO in December 1990, the full-scale single-stage-to-orbit vehicle (the DC-1) would take off and land vertically. It would be 111 feet (34 m) tall, have a base diameter of 35 feet (10.7 m), and weigh 58,005 pounds (26,366 kg) empty of fuel (669,413 pounds, 304,279 kg, with fuel). The one-third-scale *DC-X* was 39 feet (12 m) tall and weighed 22,000 pounds (10,000 kg) empty (41,630 pounds, 18,922 kg, with fuel).[72] The greatest change from the original concept was the decision to replace the aerospike engine with a conventional bell nozzle engine.

Using the aerospike engine probably would have compromised the program schedule, or at least that is what Jay Penn's team warned in the Aerospace Corporation review of the McDonnell Douglas single-stage-to-orbit vehicle. They suggested using a modified bell nozzle engine, such as the SSME (Space Shuttle Main Engine) or the RL-10 (built by Pratt & Whitney). This alternative would allow them to meet the program schedule, and they could upgrade or replace the vehicle's engine later during the flight testing portion of the program. Meanwhile, they could use the modified engine to demonstrate many of the operability features that were a desired SSTO program goal.[73]

McDonnell Douglas independently came to the same conclusion as the Aerospace review. Indeed, all Phase I contractors concluded that the aerospike engine was far too risky and that it had no performance benefits over traditional bell nozzle engine technology, a technology that the industry and aerospace engineers understood better.[74] The aerospike introduced a number of developmental and design complexities as well as program risks.[75] "The bottom line," Gaubatz recalled, "was that we could not convince ourselves that adapting the aerospike wasn't a better Ph.D. thesis than a near-term solution. The promise it held for performance was always very interesting, but the added complexity and uncertainties didn't make it a very good first choice."[76] NASA's *X-33* program later would more than demonstrate the difficulties of developing an operational aerospike engine and staying on schedule.

Instead of an aerospike, McDonnell Douglas chose a highly modified version of Pratt & Whitney's RL-10 engine, which was based on their *Centaur* upper-stage booster.[77] The choice was not just a technological decision. The firm performed extensive computational fluid dynamics analysis, as well as some wind tunnel testing, and considered results from prior studies. In addition, they could acquire the engines inexpensively. McDonnell Douglas learned that Pratt & Whitney had some available. The engine manufacturer was keen on supporting the project, so the two firms worked out an arrangement. McDonnell Douglas would borrow the engines from Pratt & Whitney, and the SDIO, through McDonnell Douglas, would pay for the engine modifications, which Pratt & Whitney's Government Engines and Space Propulsion Division in West Palm Beach, Florida, carried out.[78]

Building the rest of the *DC-X* also involved using modified existing technology. No technology development took place, because the emphasis was on operations. Some items, such as welding rods and hinges, came from local hardware stores. Deutsche Aerospace of Munich, Germany, built the landing gear, and Burt Rutan's Scaled Composites, Inc., Mojave, California, was the contractor for the exterior graphite composite aeroshell. The *DC-X* navigated with a GPS receiver and used the Honeywell avionics system built for F-18 and F-15 fighter planes, as well as a Honeywell computer. McDonnell Douglas workers salvaged and modified pressure regulators and cryogenic valves from *Thor* missiles that formerly had been positioned in Europe against the Soviet Union, saving perhaps as much as six months of labor.[79]

Curiously, the fabricator of the tanks for the liquid oxygen and hydrogen was not an aerospace firm but Chicago Bridge and Iron (CBI), located outside Birmingham, Alabama. The firm made large tanks for petrochemical works and had experience building large cylindrical aluminum containers for navy submarines. With an eye to eventually building the full-scale vehicle, McDonnell Douglas wanted a company that could manufacture a 20-foot tank, although the *DC-X* required much smaller tanks.[80] They made the tanks out of an alloy known as 2219 aluminum, although its use resulted in some welding difficulties.[81]

The *DC-X* launch site exemplified McDonnell Douglas's success in achieving SSTO program operational goals.The Flight Operations Control Center at the White Sands Missile Range, New Mexico, consisted of a compact, low-cost, 40-foot (12-m) mobile trailer that contained all necessary ground support equipment. Three people operated all the equipment and launched the *DC-X*, not the hundreds typical of rocket launches. Using liquid hydrogen, rather than kerosene or other more volatile rocket propellants, helped to simplify operations.[82] Only one person oversaw the loading of propellants, while two others set up the flight plan. Just thirty-five people formed the launch preparation and turnaround crew.

The *DC-X* design allowed it to undergo a complete turnaround in three days. It eventually demonstrated a one-day turnaround (twenty-six hours) as well as the potential for an eight-hour turnaround.[83] On-board monitoring also contributed to improve operations and maintenance. Sensors appraised the condition of critical subsystems and signaled the need to replace a component. Pete Conrad, the *DC-X* "flight manager,"[84] insisted on aircraft-like launches. He had the *DC-X* design engineers meet the firm's engineers who were designing a fully automated aircraft cockpit in order to learn more about aircraft methods.

Conrad was not new to the airplane world. After an astounding and colorful career as an astronaut, Conrad joined McDonnell Douglas in March of 1976, becoming vice president of commercial aircraft sales and later vice president for international marketing of all McDonnell Douglas products, including helicopters, fighter aircraft, and rockets. In 1990, he met Bill Gaubatz, who was managing the firm's *DC-X* work. When Conrad heard that Gaubatz's team was consulting with former astronaut John McBride, he said: "That's rather silly. . . . You obviously have the wrong guy. I'd be more than happy to help." Conrad joined the project and went on a hiatus from his other corporate duties.[85]

When McDonnell Douglas won the Phase II SDIO SSTO contract, Conrad realized that someone would have to fly the vehicle. "I remember very specifically telling Bill Gaubatz that I wanted to fly it, and be the 'pilot,'" Conrad recalled. "At that point in time, I invented the term 'flight manager' for a very specific reason. . . . I renamed myself 'flight manager,' because I was managing the system. I wasn't flying the system. The computer was flying the system." "You're going to get out to the end of the runway," he explained, "and you're going to run the engines up, and away you're going to go." After lighting the engines, flight control brought them up to 30 percent of power. If the computer said that all four engines were operating satisfactorily at 30 percent "we went to the flight mode and throttled up. It just flew off the stand."[86]

On April 3, 1993, the *DC-X* rolled out, four months ahead of schedule, on budget, and ready for flight tests. This was in itself a major aerospace achievement. Also noteworthy was the software, which the subcontractor had developed and debugged on schedule too. Nonetheless, in the twenty months that had passed since McDonnell Douglas had won the SSTO Phase II contract, the world had changed, and the SDIO with it. Those changes had a major impact on the *DC-X* test flights and on the program itself, which was essentially over before the *DC-X* began to fly. With the changing New World Order came needs for new missile defense systems and leaner federal budgets.

New World Disorder

Between the meeting with Quayle on February 15, 1989, and the rollout of the *DC-X* on April 3, 1993, four years that largely fell within President George H. Bush's incumbency, the world political order underwent dramatic changes. At the same time, the *SSX* evolved from a program and vehicle concept into the SDIO's SSTO program, then into the McDonnell Douglas *Delta Clipper.* Initially, the project seemed immune to the changing world. History and political change, however, soon caught up with the *DC-X.*

The political events that took place during Bush's term were far more momentous than those that had marred the last years of the Reagan presidency. The Strategic Defense Initiative was under attack. The revival of arms control talks coupled with Gorbachev's skillfully polished charm and salesmanship fostered a climate that was increasingly unfavorable to the SDI. A growing number of scientists, engineers, clergy, and members of the Left counted themselves as opponents of SDI. Critics of SDI and the Reagan administration had fresh fuel with the revelation in the fall of 1986 of the Irangate scandal that implicated the president.The United States had sold weapons to Khomeini's Iran (in exchange for the release of the American hostages), then diverted the proceeds to the Nicaraguan Contras, despite a congressional ban on support for the rebels. The Iran-Contra affair implicated President Reagan, Vice President Bush, the National Security Council, and clandestine, semiofficial groups on the far Right. Upon assuming the presidency, George H. Bush pardoned those involved in the affair. Meanwhile, the stock market crash of October 1987 had wiped out over 20 percent of the market's value amid revelations of insider trading and the failures of numerous savings and loan institutions.

Although it is tempting to view the Bush administration as simply a continuation of that of President Reagan, and certainly most Reagan space initiatives persisted through the Bush era, the period was one more of change than of continuity. Although the nation's civil and military space programs felt that change, the most profound changes were taking place abroad. The Reagan administration, to paraphrase Sir Winston Churchill, had fought the cold war on

Fig. 4. Rollout of the *Delta Clipper,* April 3, 1993, four months ahead of schedule. (Courtesy of the NASA History Office.)

the ground, under water, in the air, and in space. Now, the cold war was coming to an end. What use was the military-industrial complex in this postwar era? More to the point, what use was the Strategic Defense Initiative in what President Bush called the New World Order?

In retrospect, it is not clear when the long chain of events that marked the end of the cold war actually began. Certainly when Soviet General Secretary Mikhail Gorbachev addressed the United Nations General Assembly on December 8, 1988, the end was not far off. After speaking about recent domestic changes in the Soviet Union, including a restructuring of the country's political and economic life under the rubric of democratization, Gorbachev announced drastic cuts in the Soviet military presence in Eastern Europe and along the Chinese border. These cuts included a withdrawal of six tank divisions from East Germany, Czechoslovakia, and Hungary, and the entire disbanding of Soviet forces in those countries by 1991. Later that month, Gorbachev met with President Reagan and President-Elect Bush to discuss ending the cold war, but the United States remained cautious.

In March 1989, Hungarian Prime Minister Nemeth visited Moscow to tell Gorbachev personally that his country was planning free multiparty elections. In June 1989, Poland held multiparty elections. The Communists lost in both Poland and Hungary. After a summer of unrest in East Germany, in November 1989 Germans began demolishing the Berlin wall, a symbol, and a tangible reality, of the cold war. The reunification of East and West Berlin manifestly signaled the end of the cold war. Next, Czechoslovakia and Romania cast out their Communist rulers. Then in January 1990, Estonia, Latvia, and Lithuania, annexed by Stalin in 1940, demanded total independence from the USSR. All of this took place between the linking of the *SSX* to Brilliant Pebbles and the start of the SSTO program in May 1990.

By August 1991, when McDonnell Douglas won the Phase II contract, more upheaval was underway. In January 1991, after Soviet troops clashed with protesters calling for independence in Vilnius, the capital of Lithuania, Gorbachev drafted a treaty that loosened ties between Moscow and the socialist republics that made up the Soviet Union. It was due for signature in August 1991. But after Gorbachev refused a request by die-hard Communists to declare a state of emergency and hand over power, the conspirators placed him under house arrest. In the chaos that ensued in the streets of Moscow, Boris Yeltsin entered the Parliament building and prepared to resist. Yeltsin had become parliamentary leader of the Russian Republic in May 1990. On the evening of August 19, 1991, the hard-liners admitted defeat in a televised press conference.

The following month, on September 27, 1991, President Bush announced that the air force was taking all *Minuteman II* ICBMs off operational alert. In his 1991 Christmas broadcast, and later in his 1992 State of the Union address, Bush declared the struggle between the United States and the Soviet Union "now over." A new political era had begun. One of his last acts as president was to sign the START II Treaty in Moscow, which would limit the total number of each nation's warheads to 3,500 if implemented.[87] By then, though, the Soviet Union no longer existed. In Minsk, on December 8, 1991, Russia, Belarus, and Ukraine had dissolved the Soviet Union and set up the Commonwealth of Independent States (CIS). With the creation of the CIS, a new post–cold war era had begun.

As these events took place in Eastern Europe, the United States and its allies fought a war in the Middle East that radically changed the country's thinking about missile defenses. The impact on the SSTO program would be calamitous. The war against Iraq began after its annexation of Kuwait on August 8, 1990, and ended March 3, 1991. The experience of the war and Iraq's use of *Scud* missiles demonstrated to many the wisdom of shifting the emphasis of the SDI program to war-theater defense. It seemed likely America would be facing more theater wars against terrorists and renegade states rather than global nuclear war. Cold war defenses seemed no longer necessary, and programs intended to serve cold war needs no longer appeared to be appropriate.

Earlier, in late 1989, in recognition of the changes taking place in Eastern Europe, President Bush ordered a review of SDI as part of a broader reconsideration of U.S. strategic requirements for the New World Order. The review was completed in March 1990. On July 10, 1990, Henry F. Cooper became the third director of the Strategic Defense Initiative Organization. Cooper had served as America's chief negotiator at the Defense and Space Talks in Geneva since 1987, and he supported changing the emphasis of SDI. Cooper recommended that SDI be transformed to concentrate on developing defenses against limited attacks rather than preparing for an attack by thousands of Soviet warheads.[88]

Subsequently, on January 29, 1991, during Operation Desert Storm, President Bush announced that the Defense Department was shifting the emphasis of the SDI program from defense against a massive Soviet missile attack (SDS Phase I) to a system known as GPALS (Global Protection Against Limited

Strikes). GPALS had three main components: a ground-based National Missile Defense to protect the American people, a ground-based Theater Missile Defense to protect friendly nations, allies, and deployed American forces; and a Space-Based Global Defense that could stop a small attack against virtually any point on the globe.[89] The new defense architecture meant a greatly reduced version of Brilliant Pebbles. As the program underwent restructuring during 1990 and 1991, the estimated need for spacecraft interceptors fell by three fourths to only a thousand.[90]

Where did the SDIO's SSTO program fit into this new space-defense architecture? If it no longer was useful to the SDIO, could the full-scale vehicle that the SSTO program proposed to create also serve civilian or commercial space needs in the midst of changing defense and budgetary requirements? Or, was the SDIO's SSTO program simply out of place? SSTO Program Manager Pat Ladner told a reporter: "GPALS and SSTO are not tied together whatsoever."[91] If not Brilliant Pebbles or GPALS, what, then, would be the mission of the SSTO program vehicle following the end of the cold war? Perhaps the answer lay with President Bush's civilian space program.

NASA Gets a Bull in Its China Shop

Bush inherited what appeared to be an active and successful civilian space program. The space shuttle was flying again in the aftermath of the *Challenger* disaster, and NASA still was designing a space station. The country appeared to be on its way to having a single-stage-to-orbit transport, the National Aero-Space Plane (NASP). The first commercial launches licensed by the Department of Transportation were underway in 1989. The Bush administration expanded this space program on July 20, 1989, twenty years after Neil Armstrong's "giant leap for mankind," with the Space Exploration Initiative (SEI), a plan to return to the Moon, set up a lunar base, and land an astronaut on Mars. The president announced the SEI while standing on the steps of the National Air and Space Museum. Ten months later, speaking at Texas A&M University, he set the deadline for the manned landing on Mars—the year 2019.[92]

SEI would have been another large, expensive, long-term project like the space station and NASP. The initial SEI budget request was for $1.3 billion; however, NASA officials anticipated that the project's budget eventually would more than double to about $30 billion per year (unadjusted for inflation). The Senate Commerce Committee essentially zeroed out SEI funding before the program began.[93] The NASP program also came under attack in Congress in 1989. The House and Senate sought ways to cut the debt-burdened federal budget in compliance with the Gramm-Rudman-Hollings Act passed in 1985 (and modified in 1987), which mandated a balanced budget by 1991 (later changed to 1993). The mandate for these cuts predated the end of the cold war. The Economic Recovery Act and the Omnibus Budget and Reconciliation Act, both passed in 1981, combined with the massive military buildup and the costly Strategic Defense Initiative, had created an enormous federal deficit, $200 billion in 1985, which continued to grow. Meanwhile, the annual interest on the national debt climbed from $96 billion in 1981 to $216 billion in 1988.[94]

One of the most significant impacts of the Bush presidency on the space program was not a new space initiative, but a reform of NASA under a new administrator who brought "faster, cheaper, smaller" to the forefront. Ronald Reagan had supported the space agency in one of its darkest hours, the *Challenger* disaster. "We'll continue our quest in space. There will be more shuttle flights and more shuttle crews," Reagan pledged. "Nothing ends here."[95] During the Reagan years, Republican enthusiasm for NASA had a voice in Newt Gingrich, as well as in the Citizens' Advisory Council, who advocated larger NASA budgets.

Nonetheless, a different, more critical assessment of NASA arose from the ashes of *Challenger* and led to a call for reform. The official incident investigation revealed fundamental flaws in the way NASA operated. The malfunctioning Hubble Space Telescope seemed to signal further a general incompetence within the agency. Criticism grew, as did calls for change. Vice President Quayle and the National Space Council were first among those seeking to transform NASA. Congress had created the National Aeronautics and Space Council in 1958 along with NASA, but President Nixon had eliminated it. An act of Congress reestablished it as the National Space Council in 1988, effective the following year. The council's titular head was the vice president, and its specific duties involved coordinating space policy within the various agencies of the executive branch.

Quayle wanted to "shake up" NASA, which he believed was "still living off the glory it had earned in the 1960s." He complained that NASA projects were "too unimaginative, too expensive, too big, and too slow."[96] A typical example, not pointed out by Quayle, would be the Bush administration's own

Space Exploration Initiative. Although imaginative, in a certain sense, it certainly was, to use Quayle's words, "too expensive, too big, and too slow." He, like many other NASA reformers, wanted the agency to undertake "faster, cheaper, smaller" projects. If NASA shifted from large, prolonged, expensive projects to smaller, faster, cheaper projects, critics argued, the agency would be able to accomplish more science for less money.

The real struggle between Quayle and NASA, however, was over whether the White House or the space agency should determine national space policy. NASA "wanted to keep making space policy themselves," Quayle complained.[97] What lay behind the charge was a personal conflict between Quayle and the NASA administrator, Vice Adm. Richard Truly. President Bush had named Truly to the position on July 1, 1989. If any NASA administrator ever stood for the space shuttle, it was Richard Truly. He had been a shuttle astronaut on two flights: the second operational flight, November 12-14, 1981, on *Columbia,* and the eighth flight on *Challenger* from August 30 to September 5, 1983, the first flight that featured an African-American astronaut, Guion S. Bluford, Jr. Before becoming NASA administrator, Truly oversaw the heroic struggle to return the shuttle to flight and to institute measures intended to safeguard future flights. For those pushing to reform NASA, however, the space shuttle symbolized what was wrong with the agency.

The rift between Truly and NASA on one hand and the vice president and the National Space Council on the other surfaced frequently. For example, they clashed over the policies and programs to be studied by the Augustine Commission, an independent committee charged with evaluating national space policy. Known officially as the Future of the U.S. Space Program, the commission took its common name from its chair, Norman R. Augustine of Martin Marietta. Truly did not want the commission to look at the space station or the Space Exploration Initiative, which the vice president wanted reviewed. Quayle became convinced that he wanted to replace Truly. The agency, he believed, needed not a "caretaker" (Truly) but "a bull in its china shop." Eventually, acting under pressure from President Bush, Quayle, and the Space Council, Truly resigned.[98]

Quayle and the Space Council favored Daniel S. Goldin for the role of "bull" in NASA's "china shop." Quayle characterized him as "just the person we needed, and after he went to NASA he started breaking some china."[99] A former research scientist at NASA's Lewis Research Center, Goldin was then vice president and general manager of TRW's Space and Technology Group (satellite systems), where he dealt with military reconnaissance, electronic intelligence, Milstar communications, and other defense programs. He had firsthand knowledge of, and a high regard for, the SDIO and its "smaller, faster, cheaper" approach. Goldin had increased TRW work on smaller, lighter spacecraft and had introduced novel management approaches to cut program costs. As leader of his firm's participation in project-related research, Goldin was a strong proponent of Brilliant Pebbles. He also advocated cutting launch costs. In a 1991 report, he called for development of "rapid, reliable, economical" launchers out of concern that sixteen of the world's nineteen commercial launches had taken place outside the United States.[100]

Goldin became the new NASA administrator on April 1, 1992. He remained in that position through the two terms of President Clinton and into the term of President George W. Bush. When he retired in 2001, no other NASA administrator had served as many consecutive years, or under as many presidents. Goldin started the changes demanded by the reformers; he made "faster, cheaper, smaller" his own, though as "faster, better, cheaper." Quayle himself had used the longer expression "faster, safer, cheaper and better."[101] Goldin's long incumbency seemed to assure the triumph and continuing dominance of "faster, cheaper, smaller" within NASA.

Part IV / Spaceship Wars

CHAPTER 8

W(h)ither SSTO?

The installation of Dan Goldin as NASA administrator in 1992 eventually would have an impact on the SDIO's SSTO program, as would the elections that November. In the meantime, the end of the cold war and the redefinition of the Strategic Defense Initiative promised to make 1992 the first of several difficult, if not agonizing, years for the SSTO program, even as McDonnell Douglas continued to build the *DC-X*. Max Hunter's vision of a single-stage-to-orbit experimental vehicle created in a "faster, cheaper, smaller" program to demonstrate aircraft-like operations was no longer a set of ideas set out on paper and briefing charts, but a federal military program and a flight vehicle. The real challenge of achieving the vision, however, would not be creating the program or the vehicle, but dealing with the changing political arena of the immediate post–cold war era.

The SDIO budget already had come under attack in 1990 and 1991, when a Democratic-majority Congress consistently voted lower funding levels than those requested by President Bush. The SSTO program accounted for a rather small segment of the SDIO budget. Phase I amounted to only $12 million, and McDonnell Douglas received just $59 million for the *DC-X*.[1] Nonetheless, this last amount represented a budget reduction that translated into only one contractor award instead of two for Phase II.

These cuts, moreover, preceded the new post–cold war demand to reduce the Defense Department budget and reflected application of the Gramm-Rudman-Hollings Act. After years of growth under Ronald Reagan, the defense budget stabilized then declined during the Bush administration. After escalating to a high of $292 billion in fiscal 1988, military spending varied between about $290 and $291 billion during fiscal 1989 through 1991 before dropping to $276 billion in fiscal 1992 as Congress sought to collect "peace dividends." The defense budget continued to diminish even after Bush's term ended.[2] "Peace dividends" translated into announcements of defense facility closures and aerospace and defense unemployment, as the aerospace industry entered an era of dramatic restructuring and downsizing. Defense cuts also meant that military launch programs would have to compete aggressively for an even smaller slice of the shrinking pie chart that represented the defense budget.

Although the *Titan IV* expendable launcher and the *Centaur* upper-stage rocket came under congressional scrutiny, the two major launch systems competing for funds were the National Aero-Space Plane (NASP) and the Advanced Launch System (ALS). The ALS became the National Launch System, then Spacelifter, in a very short time. Its successor today is the Evolved Expendable Launch Vehicle (EELV) program. These launchers originated in an air force program called the Heavy Lift Vehicle (HLV) whose purpose was to put pieces of the Strategic Defense Initiative into space. When some members of Congress objected to the Defense Department, rather than NASA, becoming the agency most involved in future launch vehicle development, the Reagan administration proposed a joint NASA-Defense program, the Advanced Launch System, which primarily involved technology development. The intent was to make the ALS a successor to the shuttle and the *Titan IV*. One of the main focuses of ALS was NASA's development of a new engine, the Space Transportation Main Engine (STME). Then, in 1991, the Bush administration instituted the National Launch System (NLS), a joint NASA-Defense project to create a family of three heavy launchers using the STME. The largest would have serviced the space station. NASA and the Pentagon estimated the cost of the NLS to be about $11 billion, revised in 1992 to $12.2 billion. NLS defenders claimed that it would reduce launch costs, while its critics pointed to its exorbitant (and rising) price tag.[3]

Meanwhile, support for NASP waned. NASA Administrator Goldin limited his agency's annual program contribution to $75 million, while the Defense Department set its NASP funding level at no more than twice that of NASA.[4] That decision left the ALS and NLS as the primary source of hostility toward the SSTO program. Bill Gaubatz, *DC-X* program manager, found the animosity "interesting."

> There was an interesting dynamic going on at the time between the SDIO project and the NASP project office, and interestingly, with the National Launch System. It was interesting. It was a little amusing because here was this fifty million dollar program [*DC-X*] that these billion dollar programs were afraid that it was going to cut them off and that it was out to kill them. There was not a good relationship at the project levels between those programs. A lot of animosity. . . .

> I know it because people have told me. There was a lot of concern on the Air Force side for their National Launch System. . . . We had our hands full of doing what we were doing, let alone trying to work some kind of national strategy to kill these other programs. In fact, we would try to go out of our way for the NASP program, in particular, because it was true that much of what we were doing was enabled by work that had been done on the NASP program. As a result, we were not very welcome at NASA, but there was a small group at Langley that was very supporting and helped us on our tunnel testing.[5]

Aside from these funding questions, a more important question was, what mission would the full-scale *Clipper* serve? The cold war was over. President Bush had revamped the Strategic Defense Initiative. The Phase II SSTO program Statement of Work suggested various military, civilian, and commercial roles for the spaceship, including space station support and President Bush's Space Exploration Initiative. But was the role of the SDIO to develop launchers for commercial or air force or NASA use? Was the SDIO still the right institutional home for the SSTO program?

Friends of the *DC-X*

Throughout 1992 it seemed that both Congress and Defense Department bureaucrats were bent on deleting funding for the SSTO program and the *DC-X* in favor of other projects. Although both institutions had the power to support or destroy the SSTO program, only a few staff members, representatives, senators, and Pentagon bureaucrats really knew that the SSTO program existed. The number of the program's adversaries was equally small. Defending the SSTO program was always a case of a small number of people trying to make large powerful institutions comply with their wishes. Congressional staff members and the space movement played a critical role, because neither the SSTO program, nor the SDIO, nor McDonnell Douglas had money to lobby for the project.

The initial friends of the *DC-X* were, of course, those who first launched it as the *SSX* in 1989. Daniel Graham, Max Hunter, Steve Hoeser, and the friends of the *SSX* briefed a number of key agencies and individuals. Among them were members of the National Space Council (Air Force Col. Simon P. "Pete" Worden, Stewart Nozette, Edward Teller, Mark Albrecht), congressional staff (aides to Senators John Warner and Sam Nunn, for example), representatives

Fig. 5. U.S. Air Force Lt. Col. Jess Sponable (then a major) and Brig. Gen. Simon "Pete" Worden (then a colonel) at the White Sands Missile Range, with the *Delta Clipper* in the background. (Courtesy of Jess Sponable.)

from Boeing and the McDonnell Douglas Space Systems Division, and former astronaut Buzz Aldrin.[6] By chance, Bill Gaubatz, then working on NASP, was one of the McDonnell Douglas employees briefed on *SSX*.[7]

The SSTO program also counted on believers in single-stage-to-orbit transport. A central figure among those believers was Jess Sponable, who eventually joined the SSTO program as program manager during its last years. Sponable worked closely with space activist groups to build program support. A member of Jerry Pournelle's e-mail group, Sponable had "many on-line discussions" with the "rabid proponents of this kind of technology." "We tried to build a consensus," he explained, "as we moved through the DC-X program with the space activist community as to how we'd move ahead with this and what the next steps were. We tried to keep them on board and interested and supportive of what we were doing. Not every program manager does this, but I thought it was important. So, I kept Jerry Pournelle tied in with where we were going, what we were doing."[8]

From the beginning, space activist groups played a vital role as public advocates for the project. The organizations that initially promoted the *SSX* also endorsed the *DC-X*. These included Daniel Graham's High Frontier, Jerry Pournelle's Citizens' Advisory Council on National Space Policy, and the L-5 Society, where Pournelle had influence. Steve Hoeser helped to enlarge the circle of program friends by contacting the National Space Society (NSS) in 1990 "to help ensure this [program] gives us 'our' access for every American (and others)."[9] During that same year, Graham formed the Space Transportation Association in order to promote not just the *SSX*, but all single-stage-to-orbit transport.[10] The lobbying conducted by both the Space Transportation Association and High Frontier in 1990 focused broadly on all single-stage-to-orbit vehicles. Jim Muncy, a space activist in his own right and an associate of Newt Gingrich, helped to found the Space Frontier Foundation in 1988. Under the leadership of Rick Tumlinson, the Space Frontier Foundation promoted a variety of projects aimed at lowering launch costs, including the *DC-X*.[11] Henry Vanderbilt, a member of Pournelle's Citizens' Advisory Council and editor of its white paper *America: A Spacefaring Nation Again*, advanced the cause of the *DC-X* through his own space activist association starting in June 1993. Vanderbilt's Phoenix-based Space Access Society issued reports called alternatively *DC-X Updates, DC-X News*, or *DC-X Status Report*, until their renaming as *Space Access Updates* with issue 19, on September 29, 1993.[12]

Additional support came from other space groups not immediately affiliated with Pournelle or Graham, such as the National Space Society (and its lobbying arm, SpaceCause) based in Washington, D.C.; the Space Frontier Foundation; and the Aerospace States Association in Washington, D.C. These groups helped in a number of ways. For example, in February 1994, the Space Frontier Foundation, the National Space Society, the Space Transportation Association, and the California Space Development Council jointly sponsored a breakfast meeting with members of the Congressional Space Caucus, which Rep. Norman Y. Mineta and Rep. Herbert H. Bateman co-chaired.[13]

For the most part, space groups offered their newsletters, telephone trees, fax machines, manpower, and other resources in support of the congressional and bureaucratic struggles to keep the SSTO program funded. Their efforts played a vital role, at least partly because the SDIO could not lobby on behalf of its own program. As Sponable explained: "Obviously, when you wear a blue suit in the Air Force, you can't lobby Congress, and you can't force the issue, but you can certainly make sure that private citizens in this country, and contractors in this country, understand the issues, and have a common goal as far as where you're going. That's what we strove to do: keep the space activist community lined up and supportive of where we were headed."[14]

Throughout 1992, a few speeches made by Bush and Quayle passively endorsed the program. For example, during a June 19, 1992, speech before the Industrial League of Orange County in Irvine, California, President Bush, stated: "McDonnell Douglas, their SDIO-funded Delta Clipper program [SSTO] will dramatically reduce the costs of reaching into orbit. This will ensure that we lead the world's commercial aerospace industry." An astute staff member probably inserted the lines, knowing that a presidential election was five months away and that rising aerospace and defense unemployment in Southern California was a significant election issue. Quayle also endorsed the SSTO program in a speech on national space policy delivered at Vandenberg AFB on July 24, 1991. However, the vice president apparently did not feel that the program warranted any more of his effort.[15]

In contrast, support within Congress was stronger and more effective because it came from individuals whose constituencies were affected most by the SSTO program. Keeping project funding alive was "a major task," Bill Gaubatz explained. "The Congressional folks in New Mexico were major supporters. We spent a lot of time in their offices just briefing them, so they'd understand the program objectives, the major program milestone accomplishments, the funding requirements, and the national benefits."[16] The *DC-X* test flights would took place in New Mexico. The main New Mexico backer was Sen. Peter V. Domenici, a Republican member of the powerful Appropriations Committee. He believed that the program potentially could lead to development of a spaceport (and associated jobs) in New Mexico.

The strongest congressional ally of the program in the House of Representatives was Rep. Dana Rohrabacher, a conservative Republican from Southern California and a member of the House Science, Space, and Technology Committee. His Orange County constituency included Huntington Beach, Seal Beach, and a significant segment of the nation's aerospace industry. Here were the McDonnell Douglas division that was building the *DC-X* (Huntington Beach) and several firms working on NASP (Rockwell in Seal Beach) and other space-related contracts. Climbing aerospace and defense unemployment in Southern California made the *DC-X* a bread-and-butter issue for Rohrabacher, as well as for everyone representing California in Congress. Social and economic issues, not space policy, motivated voters.

Space policy, however, was also of personal interest to Rohrabacher, as was science fiction, long before his election to Congress in 1988. Around 1984, when Newt Gingrich's *Window of Opportunity* appeared in print, Rohrabacher realized that the most important space-related problem was the cost of launching payloads into space. Once he joined Congress, he supported NASP and opposed the space station because he believed that the first priority was lowering the cost of getting into space. Nonetheless, he reluctantly came to support the space station because pieces of it were being built in his district.[17]

People of the Pencils

The precarious nature of funding for the *DC-X* and Phase III of the SSTO program first became apparent in September 1991. The previous month, a power struggle in the Soviet Union saw the fall of Gorbachev and the rise of Yeltsin. The Soviet Union no longer appeared to be a menace. A power struggle of a different kind—between launcher programs—would have an immediate impact on SSTO program funding, though. This struggle illustrates the power that appointed bureaucrats wield over government spending, as well as how the alliance of politicians, staffers, space activists, and SSTO program management could work together to keep the project alive.

In September 1991, as the fiscal 1992 Defense Department authorization bill came under congressional scrutiny, the SSTO program was to be cut from the SDIO budget. Program Manager Pat Ladner telephoned the office of Rep. Dana Rohrabacher to ask for help. He was desperate. The program was dead without funding. Rohrabacher's special staff member for space matters, Tim Kyger, recalled the unusual nature of the call: "I realize now how desperate he was. Not only was he calling, but he was calling somebody in the Congress. That's something that you just don't do in that sort of position. You can get court martialed for that. They were damn desperate. They only knew about me, I guess, because of the High Frontier folks."[18]

On September 23, 1991, Kyger quickly telephoned nine members of the House Armed Services Committee. He gave them arguments in favor of retaining the program and funding Phase II at a level of no less than $30 million a year in fiscal 1992 and 1993. This would provide the money necessary to complete construction of the *Clipper*. Rohrabacher also hoped to add three sentences to the Defense Authorization Bill's conference report to shift the value of the SSTO program from the SDIO to civilian and commercial uses. "The

Congress recognizes the defense, civil, and commercial applications of the Single Stage to Orbit ('SSTO') Program," the proposed sentences read. "The exemplary 'fly-before-you-buy' development approach limits the up-front national investment, yet promises tremendous potential paybacks. The SDIO shall fund the SSTO Program at no less than $30M in FY '92 and $30M in FY '93." Rohrabacher's office also wanted Vice President Quayle to use his offices "to assure that NASA or the Air Force do not strangle this program in its 'crib.' The Vice President should not allow the NLS (New Launch System) or NASP programs to affect SSTO. . . . The Vice President should build and lead the political support within Congress and the Administration for SSTO."[19] Quayle, though, declined to lead the struggle in Congress, probably because he did not believe strongly in the program.

The real problem was not White House or congressional support, but Pentagon bureaucrats. The office of Secretary of Defense Richard Cheney sent a letter, signed by Cheney (but written by a staff member), to Dana Rohrabacher on December 2, 1991, stating that Cheney "would very much like to support the SSTO through its upcoming proof of concept phase in which a subscale, suborbital vehicle would be developed and tested." Cheney's letter pointed out Phase II: "is funded in the President's Fiscal Year 1992 budget request at $30 million, and would continue into Fiscal Year 1993. Under the $4.16 billion for SDI provided by the Congress, we should be able to continue the SSTO Program." However, the Cheney letter presciently warned, "Congressional funding reductions to the SDI program in the future could have the effect of scaling back, extending, or even terminating the SSTO Program."[20] The elimination of SSTO program funds would not come from Congress, but from Cheney's own bureaucracy.

Soon after Cheney's December 1991 letter, the Pentagon cut $27 million from the *DC-X* in order to fund the National Launch System. The budgetary maneuver utilized was a Format I. A Format I notified a program or office within the Office of the Secretary of Defense (OSD) that the OSD intended to take money already allocated to that program or office and allocate it to another program or office within the OSD. In this case, the OSD proposed to take $27 million from SDIO and transfer the funds to the NLS.

The Pentagon withdrew the Format I after the House and Senate conference on the defense appropriations bill agreed to provide $55 million for NLS out of the air force budget line, and reallocated $27 million for the *DC-X*. Still, in February 1992, the Defense Department reactivated the Format I and transferred funds to the NLS. This was the same month that the SSTO program passed a major milestone, the Initial Design Review.[21] In order to reverse the Format I a second time, Rohrabacher and other congressmen, including California's Rep. Robert Dornan, wrote to President Bush, Vice President Quayle, and Secretary Cheney.[22] Graham joined the letter writing with a message to Quayle on High Frontier letterhead.[23] Nobody could determine which Pentagon official had ordered the reactivation of the Format I in favor of the NLS. Finally, the official's position became known. A low-level bureaucrat in the Pentagon's Office of Offensive and Space Systems had decided on his own authority to reactivate the Format I. He had read a memorandum from Deputy Defense Secretary Donald Atwood on how U.S. space launch strategy should evolve, and he had become concerned that the NLS was short of funds. The final repeal of the Format I occurred on March 29, 1992; the Pentagon released *DC-X* funds to the SDIO the next month.[24]

Circus of Hells

The Format I episode illustrates how politicians, staffers, and space activists could collaborate to keep the project alive. Nonetheless, the struggle to fund the SSTO program was unending. Supporters had to maintain funding for construction of the *Clipper*, while pushing for financing of Phase III. In April 1992, Phase III funding appeared viable. Tim Kyger laid out a strategy for achieving that goal in a ten-page memorandum dated April 30, 1992. The project needed a new institutional home, he perceived, and funding for Phase III needed to start in fiscal 1994 upon completion of the *DC-X* flight tests. Phase III, in Kyger's mind, needed both the "inside game"—the "game of personal contacts" that had made possible Phase I and Phase II—and the "outside game," played by "the so-called 'space movement'"—a game of letters, phone calls, faxes, and electronic mail.[25]

For playing the "outside game," Kyger saw an array of available resources. These included the space movement, such as the L-5 Society, the Space Frontier Foundation, the National Space Society, the Space Studies Institute, and High Frontier. The Space Frontier Foundation contributed the one-page "The Space Frontier Foundation Answers Twenty Questions about the DC-X SSTO Project" (actually written by Kyger) in August 1992. The National Space Society published *Ad Astra* and had a "Space Phone Tree" (organized in 1977) that started supporting the SSTO program in the Senate in September 1992. The

Space Studies Institute contributed its newsletter, while the Citizens' Advisory Council on National Space Policy was useful for fundraising. Other resources were "computer networks," such as America Online, BIX, CompuServe, FidoNet, Genie, the internet, electronic bulletin boards, and "the science fiction community." This community comprehended such writers as Larry Niven, Jerry Pournelle, Poul Anderson, and G. Harry Stine, as well as "star trek fans" and "some members of the TNG [*Star Trek: The Next Generation* television show] cast who have gotten involved in the space movement." "Lastly," Kyger wondered, "are there any ways that the resources and influence of Hollywood (Lucas, Spielberg, etc.) can be brought to bear?"[26]

Kyger assessed the changing political climate in the Congress and the White House. "Congress next year will be an entirely New World," he wrote. The November 1992 elections, he estimated, would bring at least one hundred new representatives to Congress. "There will be new Congressional lineups on the House Armed Services Committee and its R&D Subcommittee (this is the Subcommittee that will have authority over SSTO). Ditto the House Appropriations Committee and its Defense Subcommittee. . . . Congressional leadership will be vastly different." Kyger saw this change as an opportunity to inculcate in the freshman Congress the need for the SSTO program, so he prepared a packet of materials to be distributed to each. He also foresaw presciently the need to prepare in case "there will be a new President."[27]

Kyger reflected on the impact of competing launch system programs on chances for the Phase III vehicle. NASP would be cancelled during the fiscal 1993 budget deliberations, he correctly predicted, thereby eliminating one competitor, although the program offices would not close until January 1995.[28] He believed that NLS funding had an equally dark future. "FY '93 funding for NLS is dead," Kyger wrote. "No one on the Hill—not Republicans, not Democrats, not Liberals, not Conservatives, not the DoD Authorizers in the House or the Senate, not the NASA Authorizers in the House or Senate, nor anyone on any Appropriations Subcommittee—likes or supports NLS. . . . Nonetheless, NLS is an SSTO funding threat. . . . We won't have to worry about NASP, but during the FY '93 budget exercise we *will* have to watch our backs over what the substantial NLS organization (contractor and Air Force) might do." However, Kyger felt that 1993 (fiscal 1994) "ought to be the last year NLS will be a threat."[29]

Another potential threat to Phase III funding foreseen by Kyger was the "very real chance there might be a catastrophic Shuttle accident. SSTO proponents ought also to be prepared to protect SSTO from the possible backlash of such a calamity." On the other hand, the timing of the (successful) *DC-X* test flights would be ideal for influencing the fiscal 1994 budget: "The penultimate event for the outside game SSTO campaign occurs after the November election: The DC-X flight test series. What an opportunity! These flights will be happening at precisely the time the 103rd Congress will be considering the FY '94 budget."[30]

Getting money authorized for a Phase III vehicle was still viable in spending bills. In May 1992, the final language of the Report of the Subcommittee on Research and Development (House Armed Services Committee) on the fiscal 1993 Defense Authorization Bill (H.R. 5006) authorized spending $35 million on Phase III. SDIO would spend the money on the "design, analysis, and test for further design and component development to support the development of a full scale operational prototype SSTO system." Essentially, the funds would begin development work on the Phase III vehicle. The language was adopted in a markup held on May 12, 1992. On the following day, the full House Armed Services adopted the committee report. From there, H.R. 5006 went to the floor of the House for adoption, rejection, and/or amendment. Draft language for the Senate Defense Authorization Bill (S. 2629) was distinctly different. It assumed ARPA, not SDIO, management of the SSTO program. It authorized the same spending level for the Phase III vehicle, $35 million, but in ARPA research, development, test, and evaluation funds, plus $35,676,000 to SDIO to complete Phase II, including flight testing of the *DC-X*, for a total of $70,676,000.[31]

These were authorization, not appropriation, bills. Once Congress agrees on a budget resolution, there are two funding tracks. The authorization bill lays out policy objectives and gives agencies legal authority to operate, while the appropriation bill actually spends the money. Since the 1980s, power increasingly has shifted to the appropriations bills as legislators attempted to curb the budget deficit. Generally, by moving their funding bills first, legislators passing authorization bills have tried to influence the results of appropriation bills.

This influence did not happen in the case of the House Appropriations Committee report on defense spending. It canceled the SSTO program and directed the Defense Department to assume full funding responsibility for the NLS, leaving out NASA. In addition, the ARPA research, development, test, and evaluation money requested for the SSTO was to become the SDIO contribution to the NLS program. To add insult to this injury, the Appropriations Committee report concluded that the SSTO was "not well thought out," and that the SDIO airframe designs "were hurriedly done, immature, and left technical issues unanswered." It recommended terminating the program.[32]

This was the first in a long series of attacks on the SSTO program based on the alleged technological unfeasibility of single-stage-to-orbit transport. The program itself progressed throughout June and July 1992, as the House canceled SSTO funding. The SDIO released the Environmental Assessment (EA) in June. The EA examined the environmental impact of explosions, sonic booms, toxic propellants, and the vehicle's (nonexistent) abort systems. It concluded with a favorable finding of no significant impact (FONSI). Then, between June 30 and July 2, 1992, the Final Design Review (FDR) took place at McDonnell Douglas Space Systems Company.[33] The *DC-X* was now ready to be built, but would funding be available?

In July 1992, fresh attacks seemed to come from NASA, namely, NASA's Langley Research Center via the House Appropriations Committee. A harshly critical report on the SSTO program, mistakenly believed to have been prepared at the request of the Appropriations Committee's Surveys and Investigations staff, bore the name of William Piland, chief of Langley's Space Systems Division. The staff used Piland's language to shift SSTO funding to the NLS. The basis of the Surveys and Investigations report, in fact, was largely the Langley study of Phase I contractor concepts conducted in the spring of 1991.[34] This report appeared critical of single-stage-to-orbit concepts because Pat Ladner, then SSTO program manager, had directed the Langley team to focus on the deficiencies of the vehicle concepts and design methodologies.

The attack raised a furor among the *DC-X*'s allies. On July 2, 1992, officers of the Space Frontier Foundation called for Piland's resignation for having worked with Bob Davis, staff member of the House Appropriations Committee, to eliminate funding for the SSTO program.[35] To many friends of the *DC-X* in the space movement, NASA appeared to be trying to kill the project. "We found out that a NASA official was working with Appropriations Committee staffers to kill the DC-X program," Jim Muncy explained. "I'm not trying to say that Piland's evil, but we had evidence that Piland was, in fact, working with Appropriations staffers and other staffers up here at that time to try to de-fund it. The Foundation put out a press release calling for him to be fired."[36]

According to Jess Sponable, Appropriations Committee staff members had visited him and "left impressed" with the possibilities of eventually building a single-stage-to-orbit vehicle. They also had visited Jay Penn and the Aerospace Corporation "and again thought the program was great." Next they went to Langley and Bill Piland. They saw a copy of the Phase I report, "but they weren't given the context of the report," Sponable explained. "And someone at Langley (Piland got the blame right or wrong) told them we should be studying SSTO [single-stage-to-orbit concepts], not building hardware." The "cultural clash" between "science and technology" on one hand and hardware demonstration on the other, according to Sponable, was the root cause of the Piland affair.[37]

Once these facts came to light, Rep. Dana Rohrabacher and Rep. John Murtha, a Democrat from Pennsylvania who chaired the House Defense Appropriations Subcommittee, together succeeded in restoring *DC-X* funding later in July 1992.[38] Rohrabacher commented on the episode in a *Space News* interview:

> You have very powerful interest groups at work here. You have people with preconceived notions. You have people with longstanding relationships with people who are biased one way or the other. You've got people who believe that, if the SSTO Program is successful, it will take money from programs that they have tied their careers to. . . . The fact that they had to keep this [report] secret shows their lack of confidence in their own party as to what they presented. I felt I had to go to Congressman Murtha and talk to him about specifics . . . He admitted to me that my expertise, and expertise of those of us on Science, Space and Technology [made us] better qualified to handle this question. He left it up to me to make a solid recommendation after going through the report. So I went through the report. I was not permitted to have the report. I had to read it [in an office] and do my assessment on the moment. This is just nonsense.[39]

Coming in from the Cold War

In October 1992, Congress finally approved funding for the SSTO program, but on condition that no money be spent on planning an orbital flight without congressional approval.[40] This measure did not mark the demise of Phase III, but certainly curtailed it. It did not exclude, for instance, the development of another suborbital experimental vehicle. Nonetheless, the program's institutional home was now at stake. The redefinition of the original

Strategic Defense Initiative architecture from one supported by Brilliant Pebbles to Bush's Global Protection Against Limited Strikes (GPALS) in January 1991 left the SSTO program without a clear SDI mission. The Phase II Statement of Work had attempted to trawl the waters of possible air force, NASA, and commercial uses, but the program still lacked a mission. In order for any launch system to be funded, it had to have a mission geared to the needs of an existing institution, such as NASA or the air force. Linking the *SSX* to Brilliant Pebbles originally had served that purpose. Initially, too, SDIO director Henry Cooper had supported Phase II, completion of the *DC-X*. In 1991, though, under both congressional and interagency pressure growing out of the perception that the SSTO program had become a popular rival to other launch system programs, Cooper chose not to pursue the SSTO beyond the *DC-X* (Phase II).[41]

If the project were to continue, which agency would serve as its home? Characterizing the full-scale operational DC-1 vehicle as fulfilling military, civil, and commercial missions was not the answer, although it was the truth. Launch systems suitable for a variety of missions were no longer desirable. The United States had placed all NASA, defense, and commercial payloads on the space shuttle until a tragedy showed the folly of that policy. As for the commercial launch potential of the DC-1, the country had a real need for a cheaper launcher. However, no institution for the development of commercial launchers existed, and no firm appeared ready to put up its own money, or the money of backers, to design and build the DC-1. Where, then, would the SSTO program go after the SDIO?

The question was not an entirely new one. As early as 1990, while SSTO Phase I studies were still underway, *Aviation Week & Space Technology* reported: "Few insiders expect SDIO to actually build a new generation of space launch vehicles. Even if the early work is successful and full-scale development funding available, the SSTO project would likely be passed to another organization."[42] Tim Kyger took up the question "Whither SSTO?" in a May 1992 memorandum, "SSTO Phase III: Where Should It Reside?" NASA, he ruminated, seemed to be an obvious home, "but only at first glance. NASA has an inherent conflict of interest represented by its tremendous investment in the Space Shuttle" and the NLS. "They don't, and won't, like SSTO." Furthermore, "even if NASA *did* look with favor at SSTO," Kyger wrote, "we still wouldn't want Phase III to be done there. NASA's organizational overhead would turn a $2 billion, 4 year program into a $10 billion, ten year program.

Moreover, NASA's budget is going to sustain *big* cuts each and every year for the next several years."[43]

As for the Defense Department, Kyger saw its hefty budget, "even during a massive post–World War III demobilization," as being "a large enough pot of money to have much more flexibility and 'give' than NASA's." The question, then, was which Defense agency would make the best home for the SSTO? Although the navy was interested in space, the navy would have the toughest budget battle of all the Pentagon agencies. It was busy protecting Seawolf submarine production and other projects. The army's funding fights looked as bad as those of the navy.[44] Only the air force seemed credible.

Although the air force was undergoing demobilization and downsizing, Kyger believed that: "They will be the service that will get cut the least. (That is a relative statement.)" The air force was in the middle of a reorganization that would locate the SSTO program in the Air Force Materiel Command, a new command activated on July 1, 1992, and commanded by Gen. Ronald W. Yates. Headquartered at Wright-Patterson AFB, the Command was a marriage of the former Systems Command and the Logistics Command. Part of the Air Force Materiel Command was the Space Test and Small Launch Vehicle Program Office, which Steve Hoeser had suggested as a future home for the SSTO project. That, Kyger declared, would be "the best candidate home for SSTO Phase III of all those available." Col. Pete Worden, formerly of the National Space Council and now with the SDIO, suggested the Air Force Phillips Laboratory in New Mexico as a possible future home.[45]

As for Phase II of the SSTO program (the *DC-X*), Kyger thought that the Advanced Research Projects Agency (ARPA), under the Office of the Undersecretary of Defense for Acquisition, "would seem to be the best DoD agency to conduct SSTO Phase II if SDIO will not." Kyger saw similarities between ARPA and SDIO. ARPA's "organizational overhead" was "designed to be low, in the same way that SDIO's is, and their brief is to develop risky, longer-term technologies." The agency's major problem, though, was its small budget, which would have to be "substantially increased" to accommodate Phase III. "This," Kyger warned, "would be the equivalent of a flashing neon budget sign saying, 'Yo! Over here! Cut *me!*'"[46] Nonetheless, Kyger concluded that the two candidate agencies appeared to be ARPA and the Air Force Space Test and Small Launch Vehicles Program Office at Los Angeles Air Force Base, El Segundo, California, under the direction of Col. Robert Ballard.[47]

Daniel Graham tackled the question of the program's future home in an

August 1992 white paper issued by the Space Transportation Association, "Single Stage to Orbit Vehicles: Near Term or Long Term?" Among those assisting Graham in the paper's preparation were Bill Piland, Richard Powell, and Doug Stanley (all at NASA's Langley Research Center), Max Hunter, Bill Gaubatz, Maj. Jess Sponable (SDIO), and Steve Hoeser (then with the Coleman Research Corporation).[48] In a July 1992 draft of the white paper, Graham recommended that "SSTO Phase 3 [*sic*] should be placed under management of the Air Force as an advanced technology demonstrator (X-program) and adequately funded within the USAF Technology Executive Officer structure." He also recommended transferring SDIO and NASA funds to the air force management office "commensurate with the interest of those agencies in the SSTO vehicle for SDI support and manned versions respectively."[49]

The August 1992 final version of the white paper reflected a shift in Graham's thinking to include ARPA. The "SSTO program would be best served," he wrote, "by reinstatement under SDIO management through Phase III, demonstration of the DC-Y vehicle. If this option is deemed undesirable for budgetary or mission reasons, ARPA or the Air Force should assume management. The SSTO program should be funded modestly and immediately, with funds from Departments and Agencies commensurate with the value to their missions of a successful SSTO program."[50]

The question of the SSTO program's new home remained unresolved until the following year, when Congress prepared the fiscal 1994 budget. In July 1993, the House Armed Services Committee added a provision to the Defense spending bill that moved the SSTO program to ARPA and retained the existing management team. The office of Rep. Rohrabacher had instigated the addition of this provision with the cooperation of staff member Bill Fallon in the office of Representative Dornan, who was on the House Armed Services Committee's Subcommittee on Research and Development.[51]

In September 1993, the House Armed Services Committee, with the passage of H.R. 2401, transferred the SSTO program, along with its management team, to ARPA. The Senate Armed Services Committee left the transfer of the SSTO program to ARPA in the bill's language. By Thanksgiving, the SSTO program office had begun discussions with ARPA, the Office of the Undersecretary of Defense for Acquisition, and the air force, to decide on the best organizational structure and processes for transferring the program to ARPA.[52]

The transfer, though, did not hold any real promise of salvation for the SSTO program. Opponents of the program, journalist John Cunningham pointed out, had managed to insert language in the Senate and House bills that prohibited ARPA from allowing another agency to carry out the actual work of the SSTO program. The bills directed ARPA to pursue the program itself.[53] Henry Vanderbilt, speaking as head of the Space Access Society, pointed out that ARPA had: "no great interest in actually flying rockets itself."[54] Rohrabacher also complained in a letter to the director of ARPA later that year that the agency had done nothing to date with the money appropriated in fiscal 1994 for the SSTO program.[55]

Although the SSTO program's budget line shifted from SDIO to ARPA, the project and the *DC-X* were not fated to end up at ARPA, but at another agency, the nation's civil space agency. Just as the cold war had had political and budgetary consequences for the SSTO program, the November 1992 elections would have a considerable impact, as well. Under the aegis of President William Clinton, the first president to serve after the cold war, the *DC-X* relocated from the SDIO to NASA, from the military to the civil sector. This relocation reflected the political environment of the post–cold war era, a fresh space policy, and NASA's new interest in single-stage-to-orbit concepts.

CHAPTER 9

The Disorder of Things

On January 20, 1993, Chief Justice William Rehnquist swore in William Jefferson Clinton and Albert Gore, Jr., as president and vice president. For the first time since Jimmy Carter, Democrats were in the White House. The new administration would shape military and space policy to suit its own agendas, which were certain to be different from those of its Republican predecessors. Heralding these policy changes was a new name for a defense agency scorned by liberals since its beginning. On May 13, 1993, Secretary of Defense Les Aspin announced that the SDIO henceforth would be known as the Ballistic Missile Defense Organization (BMDO).[1] At the same time, the SSTO program became the Single Stage Rocket Technology (SSRT) program.

President Clinton would not be implementing the conservative space agenda, despite appearances to the contrary. Although the SDIO survived as the BMDO, ballistic missile defense was no longer based in space, for the most part, but back on the ground. Clinton's politics, like those of President Carter, favored business and economic growth. Business, after splitting its contributions between the two parties, united behind Reagan and the Republican Party in 1980. But in 1992, businesses, especially those in the high-tech sectors that would constitute the so-called New Economy, once again began contributing to the Democratic Party.

During the Clinton presidency, the country entered a long period of post–cold war peace and economic expansion marked by the lowest levels of unemployment and inflation in over thirty years. This dramatic and historic growth stood in contrast to the dire economic conditions inherited from the Bush administration. Just as Reagan had been elected to correct the country's economic course, so Clinton rode to office on a simple slogan, "It's the economy, stupid." During the Clinton presidency, unemployment fell from 6.9 percent in 1993 to 4.0 percent in November 2000, while inflation averaged 2.5 percent, the lowest rate since the Kennedy years. Between 1992 and 2000, economic growth averaged 4.0 percent per year, compared to an average growth of only 2.8 percent between 1980 and 1992. Assisting this growth was the conservative fiscal policy of the Clinton administration made into law as the 1993 Deficit Reduction Plan, enacted without a single Republican vote, and the bipartisan Balanced Budget Agreement of 1997. These acted to cut federal spending, reduce interest payments on the national debt, and turn record budget deficits into record surpluses. The national debt had quadrupled between 1981 and 1992, and the annual budget deficit had grown to $290 billion in 1992, the largest ever.[2]

Just as President Clinton's business policies marked a shift in the Democratic Party, so the conservative space agenda had undergone changes since its initial enunciation by Newt Gingrich. The commercialization and militarization of space remained on the agenda, as did the privatization of government services (like the failed Landsat project). Gingrich, a key spokesperson for the conservative space agenda, still urged adoption of a space-based missile defense and abrogation of the 1972 ABM Treaty. His major target, though, was NASA, "a people-heavy, obsolescent bureaucracy that has got to learn a whole lot of new approaches and new techniques." The agency was no longer worthy of sustained high budgets, but should return to its trim pre-Apollo structure, he argued. Gingrich's stance reflected the shift within the conservative space agenda toward "faster, cheaper, smaller" programs, which Quayle and the National Space Council had articulated earlier, including the criticism of NASA.

Gingrich would shear NASA of its bureaucracy and reorient it to just research and development like its predecessor, NACA. He would let out to industry the remainder of NASA's functions to the maximum extent possible. The agency's overriding mission would be to reduce the cost of putting payloads into orbit. At the same time, NASA would need to foster an "entrepreneurial environment," "X" projects, and minimalist management.[3] The frontier myth also remained a keystone of conservative space thinking. Jim Muncy, Chairman of the Space Frontier Foundation, elaborated: "space is a natural extension of the Earth's frontiers, and. . .opening space to human enterprise and settlement is a unique American response to some liberals' calls for limits to growth as a rationale for ever-more-powerful statism."[4]

Much of Gingrich's restatement of the conservative space agenda harmonized with a fiscal environment that sought to reduce federal spending and the national debt. The drive to lessen federal spending seemed to bring Democrats and Republicans together. Privatization removed certain services from the federal budget, while "faster, cheaper, smaller" (NASA's "faster, better, cheaper") also curtailed government spending. The Clinton administration echoed Gingrich's calls to reform NASA, and hoped to reform the entire federal bureaucracy, as well. The politically astute Clinton administration was attuned to the enthusiasm for budget cutting, and like Gingrich and his followers, called for making government more "businesslike." President Clinton announced the National Performance Review on March 3, 1993. Later known as the National Partnership for Reinventing Government, it was an application of Total Quality Management across the range of government agencies. The goal was to make government less expensive and more efficient by cutting waste and red tape. "Reinventing government," "empowerment," and "customer service" were program buzzwords. Between reforming government by making it more "businesslike" and introducing conservative fiscal measures, President Clinton appeared to have appropriated key aspects of the traditional conservative (or at least Republican) agenda. Indeed, many of the reforms seemed to echo Ronald Reagan's cost-slashing, government-cutting, efficiency-boosting rhetoric of the 1980 campaign.

Clinton space policy specifically called for NASA, as well as all government agencies, to purchase space goods and services commercially "to the fullest extent possible," and to "privatize or commercialize its space communications operations no later than 2005." In a further echo of the Reagan era, President Clinton called for a theater missile defense capability later in the decade, a "deployment readiness program" for a national missile defense, and an advanced technology program for future missile defense.[5] He stopped short of calling for the conservative desiderata of abrogating the 1972 ABM Treaty.

Clinton shifted markedly to the Right following the congressional elections of 1994, which swept Republicans into control of the House of Representatives and which elevated Newt Gingrich to Speaker of the House. By the start of his second term, Clinton had abandoned health reform and spending measures in favor of welfare reform and balanced budgets. The Clinton slide to the Right was a strategic move that reflected the heated partisanship of the shaky *cohabitation* of the Democratic White House and the Republican House. The economic boom of the 1990s seemed to mollify concerns over the political morass of presidential ethics and an attempted impeachment of President Clinton. Meanwhile, the country as a whole also continued its slide to the Right that had begun in the 1960s.

Phase III Phased Out?

The election of a new president in November 1992 was a possibility for which Tim Kyger had foreseen the need to prepare. Now, in 1993, *DC-X* and SSRT program supporters directed their energies toward funding the program's Phase III in fiscal 1994. Debate among program personnel and supporters also focused on what the next step ought to be. Should Phase II be the *DC-Y* full-scale orbital prototype, as originally planned, or a larger suborbital vehicle known as the *SX-2*? Supporters settled on the *SX-2* for political reasons. The *SX-2*, Kyger explained, "was a politically designed vehicle in the sense that the price tag was the least amount of money that they thought they could get." The cost of the three-year project might be $300 to $400 million, which was considered at the time to be a relatively small program for the Defense Department budget.[6]

The *SX-2* would be a technology demonstrator and would expand the operations envelope beyond the *DC-X*. The next step would be the *DC-Y*, a full-scale prototype of the *Clipper*. The McDonnell Douglas version of the *SX-2* would stand 50 feet (15 m) high, about half the size of the full-scale Delta Clipper (DC-1). Its mass fraction (0.75), that is, the percentage of its mass at launch that has to be propellant, would scale directly to the mass fraction required for the DC-1 (0.91). The vehicle would use eight RL-10A-5 engines (twice as many as the *DC-X*) and burn liquid hydrogen and oxygen. Flight tests would investigate aerodynamics and reentry procedures at hypersonic speeds up to Mach 5 and at a maximum altitude of about 100 miles (160 km), where space begins, as well as an expanded regimen of ground and flight operations. McDonnell Douglas would use the same manufacturing materials needed for the DC-1. They would make the airframe and liquid hydrogen tanks out of graphite epoxy, and the liquid oxygen propellant tank out of aluminum-lithium. The thermal protection covering the aeroshell would consist mainly of titanium honeycomb panels attached to the vehicle structure in the same manner as on the Delta Clipper.[7]

Even as discussion about the next step took place among program officials and supporters, the project's friends in Congress lined up to protect Phase III funding. They wrote to the Defense Department, the Office of Management and Budget (OMB), and the House Committee on Armed Services. In February 1993, for example, at the instigation of Representative Rohrabacher's office, Sen. Pete Domenici and four other members of the New Mexico congressional delegation wrote to Les Aspin, secretary of defense, and Leon Panetta, OMB director, in support of Phase III. The New Mexico legislators urged a funding level of $50 to $100 million in fiscal 1994. The SDIO (the name change to BMDO was only months away) had budgeted nothing for the program.[8]

Joining the congressional voices and letters were lobbying groups and the space movement. They often acted at the request of Tim Kyger on Rep. Dana Rohrabacher's staff, who coordinated the defensive effort. Thus, for example, in May 1993, SpaceCause president Mark Hopkins and Daniel Graham's Space Transportation Association separately wrote to Rep. Patricia Schroeder, chair of the Subcommittee on Research and Technology, House Committee on Armed Services, in support of the program, including suggested report language.[9] Tim Kyger requested the drafting and mailing of both letters.[10] Later, in June 1993, the Aerospace States Association passed resolution 93-004 (written by Tim Kyger) in support of Phase III. It also recommended specific report language for the fiscal 1994 Defense Department authorization and appropriations bills.[11]

Finding money for Phase III in fiscal 1994 was not going to be easy. Congress still was looking for places to cut the budget. In order to fund the SSRT program, money would have to come from other defense programs. In late June 1993, Schroeder asked Rohrabacher for his help in identifying dollar-for-dollar offsets in the defense budget request to support the SSRT.[12] In response, Rohrabacher identified four potential sources of SSRT funding, including taking $75 million from the $1.2 billion cut made to intelligence programs by the House Intelligence Committee. The programs to be cut were all, according to Tim Kyger, "a bunch of programs that were felt to be next to Godliness" in order to signal to Schroeder how important SSRT funding was.[13]

The appeal appeared to work. In September 1993, the House Armed Services Committee authorized $79.88 million in fiscal 1994 to begin development of the Phase III vehicle.[14] The Senate passed the Domenici-Bingaman Amendment to the Pentagon Authorization Bill, providing funding for the SSRT program, by a vote of 66 to 33.[15] On September 1, 1993, the BMDO released a notice that it intended to issue a draft request for proposals, HQ0006-94-R-0001, for industry review and comment. "The draft RFP," the document read, "is for the design, development and test of an SX-2 Advanced

Technology Demonstrator (ATD) in support of the Single Stage Rocket Technology project. The draft RFP will be available for industry review within 15-21 days of this synopsis date."[16] In September 1993, then, it appeared that Phase III would take place. Phase III funding, though, was an illusion, like Don Quixote's thirty "monstrous giants." But the reality was not as harmless as the discovery that the giants were only windmills. By September 30, 1993, the end of the federal fiscal year, the *DC-X* flight tests halted for a lack of money. Phase II money was not forthcoming, while Phase III funds, although voted by Congress, faced an uncertain future.

The program suffered a deadly attack the following month from Terry Dawson, a member of the House Science, Space, and Technology Committee staff trained in engineering administration. On October 1, 1993, Dawson held a meeting for the staffs of the committees on science, intelligence, armed services, and appropriations in the Russell Senate Office Building to review a recent industry and NASA tour he and other staffers had taken and to discuss space launcher funding for fiscal 1994. The majority of those attending the meeting did not share the vision of single-stage-to-orbit transport. They saw it as a far-term prospect, at least fifteen years in the future, one that would require major technical advances and which would be extremely expensive. They estimated the cost of an operational vehicle to be about $15 billion.[17] Ray Williamson of the Office of Technology Assessment attended the meeting. He asserted that there was an "enormous gulf" between the *DC-X* and the technology needed for an operational vehicle. The failure of the NASP program to even approach realization of a single-stage-to-orbit design probably justified his belief. Those attending Dawson's meeting certainly did not understand that the SSRT program differed from the typical large-scale government project exemplified by NASP. Not surprisingly, the meeting ended with an agreement to draft new legislation in which single-stage-to-orbit vehicles would not figure prominently.[18]

The Dawson meeting had a devastating and immediate effect on Phase III SSRT funding. A few days later, House appropriators cut the ARPA space budget in half. The SSRT program received $40 million, about half of the $75 million recommended in the authorization bill.[19] Meanwhile, Bob Dornan, Dana Rohrabacher, and others tried to restore funding in the Senate appropriations bill, at least at the $40 million level approved by the House.[20] They succeeded somewhat. Congress proposed to give $17 million in fiscal 1994 to ARPA for the SSRT program. About $5 million would pay for completion of the *DC-X* test flights; the balance would fund the start of Phase III.[21] The final version of the appropriations bill was both more generous, giving the SSRT $50 million, and more ludicrous, failing to fund the *DC-X* test flights. When the Defense Appropriations conference bill finally went to the White House for signature on November 10, 1993, ARPA received $40 million for the SSRT program, of which $5.1 million was earmarked for finishing the *DC-X* flight tests.[22]

A great deal of congressional maneuvering and pressuring had made possible that limited success. Despite the enactment of this legislation into law, by the end of January 1994, ARPA still had not released the $5.1 million intended for the *DC-X* test flights.[23] Again, a bureaucratic procedure thwarted the will of the Congress and the friends of the *DC-X*. The procedure was a recision, a reduction of program funds that the Executive Branch proposed to Congress. On December 31, 1993, the Defense Department Comptroller submitted recision proposals amounting to $314.7 million for several programs, including $50 million from the "ARPA space program," that is, the SSRT program. "It was now obvious that the 'ARPA Space Program' didn't get on the kill list by accident," wrote Harry Stine, science fiction author and member of the Citizens' Advisory Council. "It also was quite clear that the SSTO program faced more than just cautious foot-dragging. People were now out to kill it openly."[24]

The battle over the recision lasted through most of January 1994. Rep. Newt Gingrich, then House Minority Whip, and Rep. Bob Walker joined the fight to save the SSRT program and the *DC-X* test flights. To make matters worse, Pentagon Comptroller John Hamre allowed only $10 million for the SSRT program.[25] Still, ARPA did not release even those funds to the program. Jess Sponable, in charge of what was left of the SSRT program since 1992, would have to cancel the project on February 1, 1994, unless ARPA released the funds to him. They did not.

The Bird Takes Wing

As the friends and officials of the SSRT program struggled throughout 1993 to keep the project funded, the *Delta Clipper Experimental* began its test flights. The flights were the most crucial part of the program. The *DC-X* was *not* about developing and testing technologies for single-stage-to-orbit launchers. In fact, the *DC-X* used no new hardware or technologies. The *SX-2*, on the other hand, would develop and test single-stage-to-orbit technologies.

The intention of the *DC-X* was the same as that of Max Hunter's *SSX*, namely the testing of aircraft-like operations with an experimental vehicle operated in a "faster, cheaper, smaller" program. Operations, not technology development, was at the heart of the *SSX* revolution in thinking that the *DC-X* embodied.

After the completed *Clipper* rolled out of its hangar on April 3, 1993, program manager Jess Sponable oversaw vehicle testing at the White Sands Missile Range, near Las Cruces, New Mexico. Over the following months, SSRT program and McDonnell Douglas personnel conducted hot test firings of the propulsion system in preparation for the first test flight, a short "bunny hop" that took place on August 18, 1993, at 4:43 PM MDT. The *Clipper* rose to a height of 150 feet (46 m), moved sideways about 350 feet (100 m), then descended softly during a 59-second flight that verified the vehicle's flight control systems and vertical landing capabilities.

This dispassionate description of the first flight fails to convey the excitement and drama of the actual launch circumstances. The uncertainties of both project funding and the weather troubled the *DC-X* team. Sponable recalled the conditions under which the flight took place:[26]

> We had everything prepped the night before, and we had a monsoon rain storm come through. Those monsoon rainstorms out on the desert are pretty rough. Lots of thunder, lots of lightning, lots of rain. I think we had like 70-knot winds or 90-knot winds coming through. I got a phone call at the hotel about two or three in the morning saying all heck had broken loose, that the vehicle was in danger, and that the mobile hangar, if you will, that we put around the vehicle, had broken and was about ready to impact the vehicle. So, I got up, and I wandered out there that early in the morning. It's about an hour-and-a-half drive from the hotel, so it was quite a drive by the time I got out there. Most beautiful double rainbow I've ever seen off there in the desert distance. The rain was starting to clear up by the time I got there. When I got there, the vehicle trenches, where the cryogenic lines feed in and the cryogenics are pumped into the vehicle, were literally filled to the brim with water. All of our cryogenic lines were under water.
>
> Somebody had forgotten to secure the shelter doors, which were designed to withstand that kind of 70-knot winds. Actually, it wasn't that they had forgotten. They just didn't have a procedure for securing those, because nobody was anticipating a 70-knot wind. But, we did get one, and it blew the door off the

ROGER RESSMEYER — ©1993 CORBIS. Source: A619F*1; DC-X Rocket Launch / Landing Sequence, RR006663

Fig. 6. Time sequence photo illustrating the "bunny hop" flight of the *Delta Clipper.* (Photograph copyright © Roger Ressmeyer/CORBIS.)

hinge. We ended up almost damaging the vehicle, but it never hit the vehicle. So, we got some equipment, and we hauled that thing off the vehicle. The vehicle was sitting there. Then we got fire trucks out there, and we pumped out all the trenches. Then we had to do checks. And we were trying to figure out if we could fly. The reality is, I was running out of money. It was kind of crazy to move ahead, except we had two things that were absolutely perfect: the Flight Operations Control Center and the vehicle. Both of them were in perfect condition, ready to go, but we had to get everything cleaned up.

We finally got it cleaned up. We finally checked all of the lines. The hydrogen and LOX [liquid oxygen] looked good. No water contamination. That afternoon was when we finally decided to fly. Again, it was a choice. If we couldn't get the flight off, we were likely to run out of money and never end up flying it at all. We were just really pushed. I mean, we had had slip after slip trying to get to

> a flight. So, we flew it that afternoon, and it flew successfully. We did have a fire in the nose cone, but it was minor. Some hydrogen trickled up from the engine, or wherever, and got into the lip of the nose cone. We ended up putting it out with a fire truck and a water cannon.
>
> What was equally interesting is, we had just a fantastic team down there. We then began to secure the vehicle. We worked until ten o'clock that night. Another thunderstorm came in that night. We didn't want to put the damaged shelter back over the vehicle, so we rigged a cherry picker with some metallic wires coming down off it as a catenary line to divert the lightning. We basically set that up over the vehicle. Literally, at ten o'clock at night, with lightning strikes in the near area, we had a dozen of us idiots out there, with a couple of people on each catenary line, trying to pound stakes into the ground with lightning striking everywhere. We didn't really have an option. It was tough.

The second flight was far less dramatic, but it received wide media attention. It took place on September 11, 1993, at 11:12 AM MDT. The *DC-X* attained a height of 300 feet (92 m) during a 66-second flight that tested the ascent and landing controls. During a third flight, at 10:30 AM MDT on September 30, 1993, which lasted 72 seconds, the craft reached a height of 1,214 feet (370 m), executed a 180-degree roll, and collected flight stability data.[27]

Despite these three successful flights, the *Delta Clipper Experimental* did not fly again until nine months later, in June 1994. The program simply had no more money. The project needed an additional $5 million to undertake from five to fifteen more flights in October and November 1993. The money was in the budget proposed by the Clinton administration and voted by Congress.[28] John Hamre, Pentagon comptroller, held up the funding.

It was time again for the friends of the *DC-X* to come to the rescue. The project had transferred from the BMDO to ARPA, but ARPA was not the best home for the SSRT program. Soon, though, NASA would take it over. In retrospect, the shift to a civilian organization seems logical. The project started linked to a cold war initiative; now the cold war was over. The Phase II Statement of Work indicated the commercial and civilian uses for the *Clipper*. Perhaps NASA could develop those. Such was not the thinking behind the transfer to NASA, however.

The friends of the *DC-X* desperately wanted to keep the project going, so they challenged NASA to step in. "Nobody in this Administration has had the guts to overrule these bureaucrats," declared Space Frontier Foundation chief

Rick Tumlinson, "so we're calling on NASA Administrator Goldin." Jim Muncy added: "What's happening to the DC-X is criminal, nothing less than first degree murder and grand theft."[29] Earlier, Sponable and Pete Worden, his BMDO supervisor, had been campaigning to relocate the program at NASA. The space agency, though, had no inherent interest in single-stage-to-orbit transport. NASA had abandoned the single-stage-to-orbit NASP program and had had the space shuttle since 1981. It did not appear to need a new launch system.

Needs can change. The end of the cold war allowed the White House and Congress to focus on reducing the enormous debt accumulated during the Reagan and Bush years. Spending cuts and debt reduction started in earnest under President Clinton in 1993. NASA was not immune to the cuts. The annual cost of flying and maintaining the shuttle consumed the greatest share of the NASA budget, and the shuttle accounted for nearly all of the activities and budgets of two NASA centers, the Kennedy Space Center and the Johnson Space Center. A new launcher could bring down those costs, but it also would threaten the safely ensconced shuttle bureaucracy.

The shuttle was a programmatic relic of the cold war. It exemplified the kind of expensive, large-scale, long-term project that reigned during the cold war. It was out of place in the fiscally conservative environment that followed. In that environment, "faster, cheaper, smaller" made sense. The space agency's culture had been set in the cement of the Apollo era and its cold war mission to the Moon. If NASA acquired the SSRT program, it would have to undergo a fundamental cultural change and learn how to manage "faster, cheaper, smaller" programs. That was partially why President Bush had named Dan Goldin head of NASA in 1992. The stage seemed set, then, for NASA to take on the SSRT program and manage it in the same way as the BMDO.

A Time of Turmoil

When Goldin became NASA administrator in April 1992, he was an advocate of "faster, cheaper, smaller," but not of single-stage-to-orbit transport. What changed his mind were the results of a NASA study known as Access to Space. Congress, not NASA, had mandated the study in the course of hearings on the agency's budget for fiscal 1993 (October 1, 1992, through September 30, 1993). The Subcommittee on Space, House Committee on Science, Space, and

Technology, hearings took place January through March 1992 in an atmosphere of fiscal and political uncertainty about the future of NASA's space program. In the words of the subcommittee chair, Rep. Ralph M. Hall (D-TX): "It is difficult to hold these hearings without recognizing that this is a time of great turmoil for NASA and the space program . . . We simply have to recognize that people today are less interested in going to Mars and more concerned about their trip to the grocery store."[30]

The Bush administration wanted to "freeze" NASA funding along with other discretionary spending levels. The agency's projected budget called for continuing expansion over the coming years. In addition to the third of the budget that paid for shuttle operations, NASA had several costly ongoing and new projects, such as space station *Freedom,* the Space Exploration Initiative (SEI), and a handful of space science undertakings, such as the Cassini mission to Saturn.[31] At the same time, NASA was carrying out two major launch vehicle programs with the Defense Department, the National Launch System (NLS) and the National Aero-Space Plane. NASA wanted the NLS to support space station *Freedom* and the SEI. All of these programs guaranteed that NASA's budget would not just grow, but mushroom, and Congress wanted to make cuts.

On September 24, 1992, the congressional conference committee that had been struggling with NASA's fiscal 1993 budget (H.R. 5679) issued its report. It recognized the "continuing and increasing difficulties confronting the agency and its role in the Nation's space program." These problems arose from the conflict between the desire to reduce the budget immediately as well as in future years and the number and expense of NASA missions already authorized and initiated. The report ordered NASA to undertake a study of space station *Freedom* and a study of "other national space requirements."

"This study," the committee report charged, "should be conducted in close coordination with the Department of Defense, and with other participating agencies of the National Space Council. The review should also thoroughly assess national space launch requirements, potential alternatives and strategies to address such needs, and make recommendations on improvements in the utilization of limited budgetary resources after revalidating both civilian and defense requirements. The report should be submitted to the appropriate committees of the Congress by March 31, 1993 [subsequently extended to July 1993], to permit formulation of multi-year program plans."[32]

Some of the mandated effort already had taken place the preceding year during an internal NASA review. On May 19, 1992, following a senior management meeting, Goldin ordered the creation of a set of six pairs of so-called red and blue teams. NASA normally used sets of red and blue teams to take an objective look at a given question, problem, or project. The blue teams, led by headquarters associate administrators, would find reasons and evidence for supporting a given side, while red teams, guided by center directors and deputy directors, played the role of devil's advocate. Acting Deputy Administrator Aaron Cohen was responsible for general oversight of the teams, and Assistant Deputy Administrator (and former astronaut) Charles Bolden saw to day-to-day team operations. An "integration team," led by Michael Griffin, then the associate administrator for exploration, had the task of combining the team results into a coherent whole. The study managers designated the six team areas as Aeronautics, Remote Sensing, Access to Space, Human Presence in Space, Robotic Exploration, and Moon and Mars Exploration. The agency's launcher needs came under Access to Space. Jeremiah Pearson ran the Access to Space blue team, while H. Lee Beach, Jr., Langley's deputy director, led the corresponding red team.[33]

Between May and November 1, 1992, the red and blue teams inquired into waste, mismanagement, and duplication as well as ways to pare back the NASA budget. The teams presented a preliminary report and recommendations in June 1992, and reported again in August, before submitting a final report in November. The red and blue team studies contributed to a string of "major initiatives and activities" heralded by Dan Goldin. For example, on September 17, 1992, he announced that the Access to Space blue team had recommended changes to the National Launch System program in coordination with the Defense Department. These changes included a reassessment of facilities needed for the program and acceleration of vehicle development.[34]

Access to Space

The study of launch requirements mandated by Congress in late September 1992 did not begin until after President Clinton confirmed Goldin's continued appointment at the agency. On January 7, 1993, within weeks of Clinton's inauguration, Goldin announced the start of Access to Space. The study would focus on future launch systems, analyze the launch needs of NASA, the Defense Department, and commerce, and develop various alternatives for addressing those needs. Goldin appointed Arnold D. Aldrich, associate administrator for space systems development, to head the effort. The previous administrator, Richard Truly, had entrusted Aldrich with the task of ascertaining what had gone wrong with the *Challenger* and to get the shuttle flying again.[35] Michael Griffin, associate administrator for exploration, assisted him. Before working at NASA, Griffin had headed the SDIO Technology Directorate and was responsible for starting the SSTO program. This would not be the last link between the SSTO program and Access to Space.

Aldrich and Griffin organized Access to Space through January and into mid-February 1993. In a January 11, 1993, memo to Goldin, they outlined the study. It would develop a comprehensive model of a launch system for NASA, the Defense Department, and the commercial launch industry for the period 1995 to 2030. The investigation would consider three alternatives. The first involved retaining the shuttle until 2030, while the second would replace the shuttle in 2005 with a new expendable launcher using state-of-the-art technology. The third was more daring. It would replace the shuttle in 2030 with "an unspecified . . . next-generation, advanced technology system . . . a 'leapfrog' approach, designed to capitalize on advances made in the NASP and SDI programs to achieve order-of-magnitude improvements in the cost effectiveness of space transportation."[36]

Stated simply, the premise behind Access to Space was that the cost of putting payloads into orbit was too expensive. Operations were too expensive and too complex; launchers were insufficiently reliable and safe; and the United States was insufficiently competitive in the global commercial launch market. The launch systems studied by the Access to Space teams would have to meet cost and performance criteria spelled out in advance. Aldrich charged the teams with making recommendations that were consistent with both the shrinking budget and NASA program needs. The bottom line, however, was reducing the cost of launching payloads.

Aldrich and Griffin divided launch system alternatives into three categories called options.[37] In Option 1, the United States would continue to rely primarily on the space shuttle, which would be upgraded, as well as on the current fleet of throwaway rocket launchers. The Option 1 study focused on three levels of shuttle technological improvements. Each level represented more radical change than the preceding, going from retrofit to the building of a new shuttle.

Option 2 considered four expendable launcher architectures. All four assumed that *Delta* rockets would continue to launch smaller payloads, while three assumed a replacement for the *Atlas* rocket, which lifted medium payloads into orbit. In all four cases, the Option 2 team sought alternatives to the shuttle and the *Titan* rocket, the nation's most powerful launchers. In some ways, Option 2 continued the launcher design studies started under the ALS and NLS programs.

Option 3 reflected the interest of Ivan Bekey, a senior staff member attached to the study who reported directly to Aldrich. He advocated exploring innovative vehicle concepts.[38] The Option 3 launcher would use advanced technologies and either rocket or air-breathing engines, like the NASP scramjet. In fact, the Option 3 team considered the NASP architecture. They also evaluated both single-stage and two-stage concepts. All Option 3 architectures were fully reusable.

The twenty-three–member Option 3 team consisted of NASA personnel and three representatives from the Defense Department, namely, the Air Force Space and Missiles Center, the Air Force Phillips Laboratory, and the Ballistic Missile Defense Organization. Of the twenty NASA delegates, half came from the Langley Research Center and the Marshall Space Flight Center. NASA headquarters, the Lewis Research Center, and the Johnson Research Center each sent two Option 3 members, while the Kennedy Space Center and the Dryden Flight Research Facility had one representative each.[39]

The central member of the Option 3 team was its head, Robert "Gene" Austin, who took over after Griffin left to run one of the space station *Freedom* redesign teams. Austin started out at the Marshall Space Flight Center when Wernher von Braun was still there. He had extensive experience in exotic rocket propulsion systems, including a nuclear rocket technology program that went from the conceptual stage to the testing of actual hardware for the Minerva engine, as well as ion propulsion and aerobraking systems. In 1987, while with NASA Headquarters' advanced transportation branch, Austin became involved in negotiations with the Pentagon over the Advanced Launch System program. He later worked for Griffin on the Space Exploration Initiative before joining Access to Space.[40]

A steering group periodically reviewed the work of the three study teams. Its membership included representatives from the Defense Department in compliance with the House committee mandate to conduct the study "in close coordination with the Department of Defense." The management group assessed the cost and performance results obtained by the three teams against common evaluation criteria, then selected a preferred option and recommended a plan of action.[41] In the fall of 1993, the group would brief the NASA administrator, the White House Office of Science and Technology Policy, and the Office of Management and Budget. They would report to Congress in January 1994.

The key figure responsible for making sense of the three option teams' reports, and for reaching conclusions on which to base NASA policy, was Ivan Bekey. According to Aldrich, Bekey was "probably . . . [the] strongest technical lead in managing the study. Ivan was the spark plug behind tying it all together."[42] Bekey saw himself as "more and more Arnie's system engineering guy on the whole Access to Space study . . . making sure that they were all working to apples-to-apples ground rules. . . . I collected all of the results from the three teams, and I put together a presentation that compared the results. I put them in a form in which they could be talked about. I put together the presentation for Arnie that he took to Congress and to the [NASA] Administrator, and, basically, I wrote the final report."[43]

After Aldrich and Griffin briefed Goldin in mid-February 1993 on the organization of the study, and Goldin gave his approval, work began immediately and continued into July, when the three option teams turned in their detailed four-volume "final" reports. The work of the Option 3 team turned out to be a critical step in the evolution of NASA thinking about single-stage-to-orbit transport. Just as the Have Region study had convinced Jess Sponable, the Option 3 study convinced its participants of the near-term feasibility of a single-stage-to-orbit vehicle. The catalyst for change was the prevailing influence of the BMDO's SSRT program.

The *X-2000*

Members of the Option 3 team visited the McDonnell Douglas hangar at Huntington Beach and the White Sands "launch complex." They saw the vast difference between NASA launch procedures and the SSRT program's aircraft-like operations. Instead of a typical NASA launch pad, with an army of technicians and an array of monitors and controls, they saw an elementary launch pad, a modest control center housed in a mobile trailer, and a lean launch crew. This approach was, and remains, revolutionary within the aerospace industry. They learned that one of the SSRT program's goals was to minimize the time between flights. In contrast, the shuttle required an army of thousands of technicians and several weeks (if not months) to prepare it for flight. The Option 3 team also learned about the BMDO's program management style: short-term (faster), less costly (cheaper) projects run by small teams (smaller). This was not one of NASA's costly, long-term heavily bureaucratic projects. BMDO management ideas permeated the July 1993 Option 3 report.[44]

Another pervasive theme in the same report was the importance of aircraft-like operations. "Every aspect of the program must improve operations," the report read. "Technology investment must focus on improved operations."[45] During a trip to Edwards AFB, a talk by the head of B-2 maintenance made it clear that "high tech" did not have to mean kid glove treatment.[46] The Option 3 team paid great attention to actual aircraft, such as the B-2, the SR-71 spy plane, the F-22 fighter, as well as commercial jets such as the Boeing 777. These aircraft served as "benchmarks" on which to base approaches to design, operations, and management.[47] "We were looking at operations, how people did operations efficiently, what technologies were available," Austin explained. "We had literally people from all over the country come in and brief the team on almost a weekly basis from the time we started in January of 1993 through May of 1993, when we put together our final recommendation. We had a pretty broad base."[48]

Another reason for the emphasis on operations in the report was Gary Payton. He wrote the section on operations. Payton had been with the SDIO during the earliest days of the SSTO program and had written the Phase I Statement of Work. While attending classes at the Air War College, he stayed aware of program developments and even participated in design reviews. Through Payton, the BMDO SSRT program had a direct impact on the NASA Access to Space study.

Option 3 members, led by Gene Austin, now became convinced that reusable single-stage-to-orbit rocket transport was technologically feasible. Their enthusiasm for reusable single-stage-to-orbit vehicles quickly crystallized into a concrete proposal for NASA to undertake its own single-stage-to-orbit program and to build an experimental technology demonstrator, the *X-2000* (for the program's final year of operation). The first public presentation of the *X-2000* took place on August 31, 1993, when Austin and Thomas J. "Jack" Lee, director of the Marshall Space Flight Center, briefed representatives of Rockwell International, Martin Marietta, and other aerospace firms. The briefing made it clear that the *X-2000* was a direct outgrowth of the Access to Space study and that it would not have been conceived or attempted without the Option 3 team's belief in the feasibility of single-stage-to-orbit transport.[49] Furthermore, the *X-2000* was a mirror image of Max Hunter's *SSX* vision: a single-stage-to-orbit rocket with aircraft-like operations to be developed in a "faster, cheaper, smaller" program using an "X" vehicle.

The *X-2000* had support at the highest level. According to Austin, Goldin "provided the stimulus" for the project. He wanted an experimental flight vehicle, and he wanted NASA to issue a procurement request in thirty days.[50] Some of those looking at the *X-2000* from outside the agency perceived it as an attempt to "capitalize on the excitement generated by the McDonnell Douglas *DC-X* experimental rocket," according to Ben Iannotta of *Space News*.[51] Others cynical about NASA's motives saw the *X-2000* as its way of getting into a single-stage-to-orbit program in order to kill it, thereby safeguarding the shuttle.[52]

Sponable and others working to save the *DC-X* and the SSRT program did not view the *X-2000* as a threat, but rather as a sign that NASA could be a future home for the single-stage-to-orbit program. They worked quietly to change the *X-2000* from a vehicle built exclusively by Marshall to one built by industry in a "Fast Track Managed" program.[53] "The NASA focus on building and flying an X-2000 rocket is exactly correct in my opinion," Sponable wrote in November 1993. "Whether X-2000, SX-2 or both eventually fly, they will provide a focused advancement for most of the key SSTO [single-stage-to-orbit] technologies. Properly executed, they will also provide a mechanism for pegging realistic weight growth margins on future SSTO rockets. Finally, both delay any final decision on SSTO propulsion and TPS requirements allowing (and focusing) additional ground based technology demonstrations."[54]

Curiously, Austin and his NASA colleagues shaped the *X-2000* and its program so that it somewhat resembled the *SX-2*. Both were suborbital single-stage-to-orbit technology demonstrators, but the *X-2000* was longer and had a significantly heavier weight without fuel. Also, unlike the *SX-2*, the *X-2000* had a winged body, like the shuttle. The *X-2000* would cost less than the *SX-2*, but would be ready to fly in the same amount of time, thirty-six months. NASA and the Defense Department would fund the *X-2000* jointly, thereby putting it in competition with the *SX-2* for Pentagon dollars. No chance of defense funding ever existed, though. The annual *X-2000* program cost for fiscal 1996 through 2000 was $445 million. NASA would contribute $225 million each year; the Department of Defense would supply the remainder. In contrast, the *SX-2* program would be a three-year program costing less than $100 million per year.[55]

The *X-2000* was one way Goldin could have a single-stage-to-orbit program. Meanwhile, the *DC-X* had been grounded since September 30, 1993, for lack of money. The Defense Department was reviewing its launch needs, and the White House was formulating a space policy that would determine new roles for NASA and the Pentagon in future launcher development. January 1994 arrived. The *DC-X* remained grounded. NASA reported the Access to Space results to Congress and released them to the public. Neither the future of the *DC-X* or the *SX-2* nor the future of the nation's launch policy were apparent.

CHAPTER 10

The *Clipper Graham*

The transfer of the SSRT program and the *DC-X* to a civilian agency following the end of the cold war raised a critical question: what would happen to Max Hunter's vision of a single-stage-to-orbit vehicle with aircraft-like operations built and tested in a "faster, cheaper, smaller" program? Although Access to Space had made NASA more receptive to instituting its own single-stage-to-orbit program, the agency might not run it in the same "faster, smaller, cheaper" way as the BMDO. Even though Administrator Goldin espoused "faster, better, cheaper," NASA still had the shuttle and a long cultural history of large-scale, long-term, expensive projects. The agency might slight aircraft-like operations in favor of technology development. In fact, technology development *was* how the agency viewed the opportunity to work on a reusable single-stage-to-orbit vehicle.

Access to Space had spawned just such a technology development program. Flight tests would take place on an experimental vehicle, the *X-2000*, while ground-based investigations would research various technologies. Ultimately, NASA hoped to build a single-stage-to-orbit replacement for the shuttle, that is, the vehicle preferred by the Option 3 team. Meanwhile, President Clinton had not yet issued any launcher or space policy, although a statement was due in 1994. NASA took advantage of the lack of direction to formulate plans for a single-stage-to-orbit shuttle replacement and courted the Pentagon as a partner. It was in the midst of this policy disorder that NASA and the fate of the *DC-X*, mothballed in September 1993 for a lack of funds, came together.

A Plan for Technology Development

Out of Access to Space came a concrete plan for achieving a full-scale, reusable single-stage-to-orbit rocket to fulfill the launcher needs of NASA, the military, and commerce. Both the July 1993 and the January 1994 Option 3 reports spelled out that plan, which incorporated existing launchers. The vehicle would service the space station and put into orbit NASA, Defense Department, and commercial payloads weighing up to 20,000 or even 25,000 pounds. Until that single-stage-to-orbit shuttle replacement became available, the government would upgrade existing *Delta, Atlas,* and *Titan* launchers between 2000 and 2008 in an "interim expendable launch vehicle program." Both throwaway rockets and the shuttle would "have to be operated for at least another 10 to 15 years before new launch vehicles can be available." Later, the single-stage-to-orbit transport would capture "Delta, Atlas, and Shuttle missions at approximately 15% of the current combined annual operating costs of these systems."[1] It was a politically cautious plan that avoided attacking NASP or the NLS so that the program would not threaten any existing NASA or Defense Department project. Indeed, the January 1994 Summary Report also recommended maintaining the NASP technology program.[2]

The major argument in favor of the single-stage-to-orbit project was that it would lower launch costs more than any "number of beneficial improvements to the Shuttle system." Design, development, test, and evaluation costs were smaller, although not in the near term, for the improved throwaway rockets. The same costs for the single-stage-to-orbit architectures were larger, but would be delayed four to five years as a result of the need to research and develop the required technologies. The operational single-stage-to-orbit vehicle would feature reduced annual costs and lower total "life-cycle" costs (the sum of all costs from design to finished craft). Developing single-stage-to-orbit technologies also promised spin-offs "with dual-use in industry (such as composite vehicle structures for cars and airplanes)." The single-stage-to-orbit craft itself "would place the U.S. in an extremely advantageous position with respect to international competition, and would leapfrog the U.S. into a next-generation launch capability."[3]

The Option 3 reports also laid out a game plan for developing the single-stage-to-orbit transport. Rather than immediately select and develop a particular architecture, NASA would initiate "comprehensive vehicle design trade studies" concurrent with "an associated technology maturation program . . . utiliz[ing] a fast-track management approach that would involve the development of flight experiments and experimental, or 'X', vehicles."[4] Here was a NASA version of the *SSX* vision: a single-stage-to-orbit craft developed with an "X" vehicle in a "faster, cheaper, smaller" program. All it lacked was aircraft-like operations.

Trials of rapid turnaround and other aircraft-like operations would take place on the experimental vehicle. The Option 3 reports also laid out a ground-based technology development program. The technology assessment subteam, under Terry F. Greenwood, Marshall Space Flight Center, identified six fundamental technologies for development: 1) reusable cryogenic propellant tanks; 2) reusable liquid oxygen–liquid hydrogen propulsion; 3) vehicle health management (hardware and software for auditing vehicle systems); 4) durable thermal protection systems; 5) adaptive autonomous vehicle software and avionics; and 6) operable vehicle subsystems. The January 1994 report gave the list a somewhat more concrete form: "graphite-composite reusable primary structures, aluminum-lithium and graphite-composite reusable cryogenic propellant tanks, tripropellant or LOX-hydrogen engines designed for robustness and operability, low-maintenance integral or standoff thermal protection systems, autonomous flight control, vehicle health monitoring, and a number of operations-enhancing technologies."[5] Most of these technologies related to aircraft-like operations either directly, such as vehicle health monitoring, or indirectly, such as the reusable propellant tanks.

According to an October 1993 briefing given to the Office of Management and Budget, the proposed technology development project, known as the Access to Space Technology / Demonstration Program, would run from fiscal 1994 through 2000. Both NASA and the Defense Department would fund the project. Its total cost would be about $2.637 billion, with all but the first two years funded at $450 million annually.[6] The proposed cost-sharing with the Defense Department mirrored the ongoing collaboration between the two agencies to develop the NASP and NLS launchers. Many *DC-X* supporters hoped to see either a joint NASA-Defense program or two competing NASA and Defense single-stage-to-orbit programs.[7]

The Moorman Report

Meanwhile, in April and May 1993, while discussing possible joint projects with NASA, the Pentagon carried out a sweeping survey of launch requirements that considered both expendable and reusable launchers.[8] Next, John Deutch, then undersecretary of defense for acquisition and technology, led a six-month, bottom-up review of spending priorities that ended in early September 1993. It considered expendable launchers and a reusable single-stage rocket. However, after the Deutch review, Secretary of Defense Les Aspin ruled out developing a new rocket. In November 1993, the White House asked congressional staff members to keep the major launch system programs alive until the administration had decided how to proceed.[9]

Throughout October and November 1993, following Aspin's decision, NASA and Pentagon officials discussed cooperation on launcher projects, including a single-stage-to-orbit rocket. As early as July 1993, Air Force Chief of Staff Gen. Merrill McPeak had written a memorandum to two Defense Department officials regarding the need to cooperate with NASA on launch and propulsion technology.[10] NASA viewed Defense collaboration as a necessity in light of current budget reductions, according to Michael Lyons, then NASA's deputy associate administrator for flight systems. However, he believed, the Pentagon did not want to undertake another joint program, so NASA proposed a cooperative arrangement that would have one agency lead the program and the other provide technical support.[11]

Disagreement between the two agencies also extended to how big a new launcher would need to be. The air force wanted one with a payload capacity of about 20,000 pounds (approximately 9,000 kg) to low orbit by 2005. That was the weight class of the vehicles investigated in the Have Region study and the SSRT program. NASA, in contrast, desired a vehicle large enough to carry 45,000 pounds (20,400 kg) to orbit in order to service the space station. NASA correctly conjectured that a single-stage-to-orbit rocket was unlikely to be ready by 2005[12] and that two launch programs would be unaffordable in the current budget environment.[13]

In late 1993, as these discussions were taking place, Congress mandated that the Defense Department conduct a new review of its launch needs. The Fiscal 1994 Defense Authorization Act directed the secretary of defense to develop a plan to establish priorities, goals, and milestones for the modernization of space launch capabilities. Lt. Gen. Thomas S. Moorman, Jr., vice commander of Air Force Space Command, led the study, which the Pentagon was to deliver to Congress on April 1, 1994.[14] Because of the so-called Moorman study, the Defense Department became reluctant to begin a new program without a plan to continue it in the years to come. Consequently, it did not wish to proceed with SSRT Phase III until the completion of the study.[15]

The Moorman report, officially titled the "Department of Defense Space Launch Modernization Plan," did not reach Congress until May 1994. It considered four launch system options, one of which was a reusable rocket.[16] NASA's Arnold Aldrich, who served on the Moorman commission, described the extensive range of testimony that went into the commission's report.

Moorman, he recalled, "had anybody in the country that wanted to talk to him about transportation or propulsion or rockets come in and tell him and the Senior Board. I bet we listened to 80 different companies."[17]

Following receipt of the Moorman report, the White House Office of Science and Technology Policy (OSTP) announced the National Space Transportation Policy. (Clinton had dissolved the National Space Council; the formulation of space policy now fell to the OSTP.) The National Space Transportation Policy had been months in preparation by an interagency group reviewing launch policy headed by Richard DalBello of the OSTP. Already, in April 1994, the White House had let it be known that it had ruled out development of a big new rocket "for the foreseeable future."[18] Moreover, the White House did not anticipate a decision on whether to pursue a new launch system until 1995.[19] In May 1994, while the National Space Transportation Policy was still in the draft stage, the OSTP designated NASA as the lead agency for reusable launch systems, and the Defense Department as the lead agency for expendable launch systems.[20] This designation later became part of the National Space Transportation Policy signed by President Clinton on August 5, 1994, and scheduled for implementation by NASA on November 5, 1994.[21]

Clinton space policy facilitated transfer of ARPA funds to BMDO for the resumption of the *DC-X* test flights. However, it also meant that the SSRT program could no longer reside with the Pentagon, but had to shift over to NASA, which now had responsibility for the development of reusable rockets. The policy angered supporters of the SSRT program. For many, NASA was not a favorite agency, and it was certainly not high on their list of agencies to provide the program a new institutional home. Representative Rohrabacher protested in a May 17, 1994, letter to Vice President Gore. NASA was "preoccupied" with the space station and other large-scale missions, he wrote. "I am loath to add yet another lead responsibility onto NASA at this time, even when it's important to NASA and to future U.S. space capabilities." The agency lacked the money to sustain the project and was unlikely to have it in the future. The Department of Defense "will at least have *some* dollars available for such allocation."[22]

Rohrabacher and others in Congress disregarded the OSTP policy. In July 1994, for example, defense appropriators in the House voted to award $50 million to the SSRT program for fiscal 1995, but only if the White House did not select NASA to run it.[23] In August 1994, both the House and Senate defense authorizers agreed to spend $30 million on the program.[24] The Pentagon, however, was not going to disburse any of the money allocated by Congress.[25] On November 25, 1994, Defense Secretary William J. Perry informed Sen. Robert Byrd and Sen. Mark Hatfield that the Pentagon was releasing $35 million from its research, development, test, and evaluation appropriation to NASA for the SSRT program.[26] For its part, NASA now had White House approval to proceed with plans to develop a single-stage-to-orbit launcher.

Testing Technology

Putting the SSRT program in NASA's hands raised serious questions about how it would be run. Whether as the *SSX*, the SDIO's SSTO program, the BMDO's SSRT program, or as the *DC-X*, the vision had been about testing the application of aircraft operations to reusable rockets. The emphasis was not on developing new technologies. NASA, however, wanted to develop technology; it was part of the agency's culture. Like its institutional predecessor, the National Advisory Committee for Aeronautics (NACA), NASA had a long tradition of developing and testing aerospace technologies. Not surprisingly, then, the agency looked for opportunities to initiate new technology development programs. That was the goal of the Access to Space Technology / Demonstration Program.

With backing from Dan Goldin, the Access to Space Technology / Demonstration Program quickly took off. The program started in Goldin's office and began to take shape. In January 1994, Goldin appointed Jack Lee as his special assistant for Access to Space. Even before Lee transferred to headquarters from the Marshall Space Flight Center, Goldin considered him to be critical to the space agency's launch technology development efforts. "We definitely need to address future launch issues and keep the heat on," he told the November 29, 1993, meeting of senior staff and NASA center directors. "Aldrich [as head of the Office of Space Systems Development] and Lee are leading this effort—which could result in a national leadership role for us."[27] Lee started his professional career in 1958 as an aeronautical research engineer with the U.S. Army Ballistic Missile Agency at Redstone Arsenal, Alabama. In 1960, he transferred to the Marshall Space Flight Center as a systems engineer and became technical assistant to the technical deputy director of Marshall in 1969. After managing various project offices, including Spacelab, Lee served as deputy director of Marshall from December 1980 to July 1989, when he became its director.[28]

At headquarters, Lee coordinated and refined the Access to Space Technology / Demonstration Program, which would take place entirely at Marshall Space Flight Center. His headquarters efforts continued those Lee had been carrying out at Marshall during the autumn of 1993. In a memorandum to Goldin dated October 8, 1993, he identified the three "key technologies . . . essential for development of a single stage to orbit (SSTO) launch system." These were reusable cryogenic fuel tanks, graphite composite structures, and a tripropellant rocket engine.[29] They became the principal technologies studied in the ground portion of the Access to Space Technology / Demonstration Program.

Lee further elaborated on the three technologies. The reusable cryogenic tanks would be made of either aluminum-lithium or graphite composite. They would last a long time and require minimal maintenance.[30] The greatest advantage of aluminum-lithium over conventional aluminum in cryogenic tank construction was a saving of 15 percent in mass. There were some disadvantages, though. For example, aluminum-lithium was about five times more expensive, had a tendency to delaminate, and required special shielding for welding. All candidate aluminum-lithium alloys, known commercially as Alcoa 2090, Alcan 8090, and Weldalite (Reynolds) 2195, already were in use in such aerospace applications as the *Atlas-Centaur* payload adapter (Alcoa 2090), the *Titan IV* payload adapter (Alcan 8090), the Airbus A330/A340 leading edges (Alcoa 2090 and Alcan 8090), the McDonnell Douglas C-17 fuselage, cargo door, and deck (Alcoa 2090), and the Westland Agusta cabin frame and fuselage (Alcan 8090 and Alcoa 2090). Lee believed the optimal tank alloy to be Weldalite 2195.

In addition, Lee was considering carbon fiber reinforced plastic (CFRP, a composite material) tanks. CFRP would save 25 percent in mass over conventional aluminum tank structures. It had disadvantages, though. Those who worked with CFRP knew that its permeability to hydrogen might increase with each refilling. However, experience with CFRP structures in contact with cryogenic hydrogen was limited. Experts foresaw problems sealing end domes, access doors, and the like. Lee explained to Goldin that primary structures, such as aeroshell sections, made of graphite composite materials would reduce overall structural weight by 40 percent compared to the shuttle structure, and without experiencing corrosion or fatigue problems. Actual transport aircraft with composite primary structures included the Boeing 777, the Airbus A330 and A340, and the ATR 72.[31]

Lee also wanted NASA to study tripropellant rocket engines and an upgraded Space Shuttle Main Engine for use on a single-stage-to-orbit vehicle.[32] The specific tripropellant engine that he wanted studied was the RD-701, which was under development in Russia. His consideration of the RD-701 signified further NASA collaboration with Russia following the end of the cold war, although not as extensive as cooperation would be on the international space station, which NASA announced earlier in 1993. The Russian RD-701 initially used kerosene and liquid oxygen, then shifted to liquid hydrogen and oxygen. Nobody, however, had yet test fired the engine. "It's still got a ways to go to be a real engine," explained Gene Austin.[33] Critical problems that research would have to address included the tripropellant injector and the mechanism for shifting from kerosene to liquid hydrogen fuels.[34]

More concrete details of the proposed Access to Space Technology / Demonstration Program came to light on October 19, 1993, when NASA officials briefed their counterparts from the Pentagon's SSTO Review Team on Access to Space. At that meeting, Dick Powell, Langley Research Center, discussed a single-stage-to-orbit vehicle concept, the *SSV (Single-Stage Vehicle)*, that his group was studying. The *SSV* featured a bonded blanket thermal protection system, graphite composite wings and other internal and external structures, and aluminum-lithium tanks for the liquid hydrogen and oxygen. The proposed propulsion system would be either seven improved Space Shuttle Main Engines or three RD-701 engines.[35]

When Gene Austin briefed the same group on the proposed technology development program, he discussed flight testing technologies on the *X-2000* Advanced Technology Demonstrator.[36] But NASA never built the *X-2000*. The project ran counter to Clinton space policy. As Austin and his Marshall colleagues envisioned the program, the Marshall Space Flight Center would design and build the *X-2000*. Industry would play a role only as a contractor to Marshall, with no input to the design of the vehicle. Goldin rejected the proposal because it failed to give industry a larger partnership role. The draft National Space Transportation Strategy, released April 26, 1994, directed NASA to use "private sector partnerships" in implementing the strategy and to "actively consider these arrangements in their decisions on technology and development investments."[37]

Goldin wrote to John Gibbons, assistant to the president for science and technology, in September 1993 that he did not want a new launcher, just a technology program. "NASA does not have a plan to develop or release any request for proposal for any launch vehicle, even as a technology demonstration. We do have a desire, however, to ensure NASA is pursuing the technology needed by the country in support of future launch vehicles." "With this in mind," he continued, "it would be desirable to identify a wedge within NASA's budget for a technology development effort that would be supportive of a program that might be defined by the Working Group [on Space Transportation of the White House Office of Science and Technology Policy] in concert with all affected agencies."[38] If NASA would not build the *X-2000* or "any launch vehicle, even as a technology demonstration launcher," how would the Access to Space Technology / Demonstration Program flight test single-stage-to-orbit technologies or, if it wished, aircraft-like operations?

NASA Eyes the *DC-X*

The answer was the *DC-X*. The Access to Space Option 3 team already had sketched plans to use it as a technology and operations testbed as early as the summer of 1993. The range of technologies they hoped to test included structures and materials (e.g., reusable cryogenic tanks), thermal protection systems, and avionics. The Option 3 team also wanted the Defense Department to pay for the remainder of the *DC-X* flights ($5 million).[39] The *Delta Clipper* had not flown since September 30, 1993, because of a lack of funds. The Pentagon bureaucracy had suspended funding, in spite of Congress.

ARPA still had not released *DC-X* funds as January 1994 came to a close.[40] Sponable would have to cancel the *DC-X* contract on February 1, 1994, unless he received those funds. He did not. Instead, on January 31, 1994, Dan Goldin released $900,000 in NASA money to keep the *DC-X* at the White Sands Missile Range and to defray the site overhead charges until ARPA released the appropriated funds.[41] Although Goldin's disbursement preceded any official White House statement on launcher policy, he surely must have had at least the tacit approval of the Clinton administration. When ARPA finally released the $5.1 million to BMDO in late April 1994, about $3.5 million went to McDonnell Douglas, while the rest covered range expenses at White Sands Missile Range.[42] "The amount of time and energy that went into getting $5 million released to continue the project," Gaubatz reflected, "was just phenomenal."[43]

At the same time as the *DC-X* rescue, NASA and the BMDO began to draw up plans for NASA to use the vehicle as a technology testbed. Heading the

DC-X evaluation team were Gene Austin and Jim Kennedy from NASA and Gary Payton, Jess Sponable, and Curtis McNeal from BMDO. Terry Greenwood led the technology team, while Marshall's Steve Cook (a former Option 3 team member) headed the *DC-X* integration and demonstration team. His team considered cryogenic tanks, primary structures, thermal protection systems, propulsion, and avionics, as well as contract and cost issues.[44]

The *DC-X* evaluation team briefed Goldin on March 1, 1994. They discussed the technologies needed for single-stage-to-orbit flight "in a Real World, Operational Environment." The candidate technologies were thermal protection panels, a graphite-composite liquid hydrogen tank, a graphite-composite structure between the fuel tanks (the intertank structure), a liquid-to-gas rocket control system, hazardous gas detection equipment, and graphite-composite feedlines and Russian flange hardware for the propulsion system. They did not leave out operations. They explained the need to perform the rotational maneuver that was critical to proving the concept of a vertically landing rocket. The *DC-X* had not yet performed this maneuver. The full-scale version of the *Delta Clipper* would return from space nose first, then perform the maneuver (turning 180 degrees), in order to land on its base, not its nose. The team also proposed managing the project in a way that was new to NASA. The "New Way of Doing Business Within NASA" included "Employee Empowerment," "Limited Government Oversight," a small budget, limited personnel, the "acceptance of Limited Risk," and the "By-Pass[ing of] Bureaucracy."[45]

These were characteristics of a "faster, cheaper, smaller" program. Goldin had firsthand knowledge of, and a high regard for, the "faster, cheaper, smaller" style of the SDIO (and the BMDO). In addition, the Clinton administration's National Space Transportation Policy enjoined NASA to "involve the private sector in the design and development of space transportation capabilities and encourage private sector financing" and to "emphasize procurement strategies that are based on the use of commercial U.S. space transportation products and services."[46]

One specifically proposed "New Way of Doing Business Within NASA" was the cooperative agreement.[47] The National Performance Review directed NASA to extend the use of cooperative agreements to for-profit organizations,[48] while the National Space Transportation Strategy urged NASA to use them to foster technological development in collaboration with industry.[49] Traditionally, NASA used (and continues to use) cooperative agreements only with nonprofit institutions and universities in cases requiring a close working relationship.[50] As a procurement instrument, the cooperative agreement allowed NASA to expedite acquisition procedures, by-pass bureaucracy, reduce project personnel, limit oversight, and shrink budgets.[51] Specifically, the agreement allowed the agency to avoid normal procurement statutes known as Federal Acquisition Regulations (FARs),[52] thereby reducing both paperwork and applicable federal regulations and accelerating the procurement process. Another important aspect is that they required industry to share project costs. The cooperative agreement was central to NASA's plan to turn the *DC-X* into its own technology demonstrator.

The plan would use the *DC-X* (which NASA renamed the *DC-XA,* where "A" stood for Advanced) as a technology testbed and a "management showcase."[53] It could not have come at a better time for the vehicle. The *Clipper* flew for the fourth time on June 20, 1994, after spending nine months in storage. Being flown after sitting in storage for so long was surely an impressive aerospace first unintended by those running the program. The flight also was the longest (136 seconds) and highest (2,600 feet, 870 meters) to date. The *DC-X* operated with a full load of propellant for the first time, and its radar altimeter sat in the control loop as part of its internal autonomous flight programming. It flew again a week later on June 27, 1994. The turnaround time of one week between flights was unprecedented and a milestone operational achievement. The flight also demonstrated the craft's ability to land under total control of internal programming. That demonstration had not been planned, however.

A ground equipment explosion caused a shock wave that ripped a 4-by-15-foot (1.2-by-4.6-m) hole in the *DC-X* aeroshell. Pete Conrad, watching his computer screen in the Flight Control Center, heard BMDO consultant, Jim French, who was standing behind him, shout: "Parts are falling off the bird!" Conrad reacted with lightning speed and activated the software command for "autoland." The *Clipper* safely returned to the ground. The computer executed a successful intact abort.[54] This maneuver was another remarkable aerospace first. McDonnell Douglas and the BMDO later determined that gaseous oxygen, gaseous hydrogen, and water vapor had been drawn into the air purge duct of nearby ground equipment, instead of being blown away. The hydrogen tank had a two-inch crack. A new aeroshell would cost $700,000.[55] However, they already had spent half the vehicle's $5.1 million budget. The project simply did not have enough money to make repairs and fly again. In fact, the damaged *DC-X* would not fly again until May 16, 1995.

A Vision Transformed

Following the accident, the McDonnell Douglas *Delta Clipper* team had to decide what to do with the vehicle. During the July 1994 progress and status report, Paul Klevatt, director of McDonnell Douglas's SSRT effort, tackled that difficult question. The good news was that they could repair the damaged composite aeroshell, liquid hydrogen tank, and other parts. There were two options, he explained. Option 1 was to repair the vehicle and complete the remaining three or four flights as soon as possible. Option 2 was to transfer everything to NASA in accordance with its *DC-XA* planning. If they pursued Option 2, they could follow two paths. Option 2(a) was to repair the *DC-X* incorporating fixes learned from the incident and to complete the flight schedule in 1995. Option 2(b) was to repair the *DC-X* incorporating the *DC-XA* changes and to fly the upgraded vehicle in 1996, as called for in NASA's plans.[56]

Ultimately, ARPA paid McDonnell Douglas $2.5 million to repair the vehicle's hydrogen tank and to replace the damaged sections of the composite aeroshell. In June 1994, NASA and the Pentagon announced that each would contribute $1 million to transfer the *DC-X* to NASA in July of 1995. After NASA took over the project, it spent $13 million in fiscal 1995 and $16 million in fiscal 1996 to fly the *DC-XA*.[57]

Upon completion of the repairs, the *Delta Clipper* flew three more times—on May 16, June 12, and July 7, 1995—before its reassignment to NASA. These three flights expanded the flight envelope by reaching increasingly higher altitudes—4,363, 5,708, and 8,202 feet (1,330, 1,740, and 2,500 m), respectively, higher than the 2,600 feet (870 m) achieved on June 20, 1994. The three flights lasted 124, 132, and 124 seconds, respectively. They did not break the record length set during the June 20, 1994, flight, however. Some of the operational goals achieved included flying with a constant angle of attack (May 16), the first use of the reaction control system thrusters (June 12), and the rotational maneuver (July 7). On the July 7, 1995, flight, the aeroshell cracked during the vehicle's rapid (14-ft/sec or 4.3-m/sec) landing. The *Clipper*'s fast descent resulted from an invalid data problem with the radar altimeter inserted in the control loop as part of its internal autonomous flight programming, not from the landing control software. Such mishaps were bound to happen to an experimental vehicle in a program that pushed operational and managerial limits.

Before those 1995 flights took place, however, NASA started the process of transforming the *DC-X* into a technology demonstrator with the issuance of two NASA research announcements (NRAs) and their publication in *Commerce Business Daily* on February 7, 1994. NASA released the final NRAs on February 23, 1994, and made an award March 3, 1994. A separate agreement or contract was issued for integrating the technologies on the vehicle. NRA 8-11, Advanced Propulsion, sought to have industry undertake research in five technical areas, including bipropellant and tripropellant thrusters and thrust cells, oxygen-rich turbine drives, and the use of Russian hardware for both expendable launchers and advanced launchers.[58] NRA 8-12, Advanced Structures and Thermal Protection Systems, focused on fuel tanks, thermal protection systems, graphite-composite primary structures, and health monitoring systems and other nondestructive means of evaluating vehicle systems.[59]

The industry briefing for the two NRAs, held on February 18, 1994, emphasized the agency's commitment to "fast-track" procurement and cooperation with industry. One overview, echoing Bill Clinton during the presidential campaign, succinctly summed up NASA's thinking: "It's Ops [operations] Cost, Stupid."[60] NASA also made it clear in other briefings that it saw a future full-scale, single-stage-to-orbit rocket as vital to maintaining the space station.[61] As a result of the issuance of NRAs 8-11 and 8-12, NASA awarded six cooperative agreements to industry (McDonnell Douglas and Rockwell) in June and July of 1994.[62] McDonnell Douglas and Rockwell worked in conjunction with NASA laboratories, such as the Marshal Space Flight Center and the Ames Research Center, as part of the cooperative arrangements. Ultimately, from the work conducted under these agreements, including a cooperative agreement with McDonnell Douglas for the integration of technologies, the upgraded *DC-XA* emerged.

The twenty-eight-month agreement between McDonnell Douglas and Marshall to reconfigure the *DC-X* called for $17.6 million in NASA funding and $7.6 million from McDonnell Douglas.[63] The specific upgrades that turned the *Delta Clipper* into the *DC-XA* were: (1) a switch from an aluminum oxygen tank to a Russian-built aluminum-lithium alloy cryogenic oxygen tank with external insulation; (2) an exchange of the aluminum cryogenic hydrogen tank for a graphite-epoxy composite liquid hydrogen tank with a low-density reinforced internal insulation; (3) a graphite-epoxy composite intertank structure; (4) a graphite-epoxy composite feedline and valve assembly; (5) a gaseous hydrogen and oxygen auxiliary power unit to drive the hydraulic systems; and (6) an auxiliary propulsion system for converting liquid hydrogen into gaseous hydrogen for use by the vehicle's reaction control system.[64]

The *Clipper Graham* Flies

The completed *DC-XA* rolled out on schedule on March 15, 1996, and arrived at White Sands, New Mexico, seven days later. After engine tests on May 4 and 7, 1996, the vehicle was ready once again to test technologies and aircraft-like operational goals. A series of five flights would: (1) verify the functioning and operational suitability of the new hardware; (2) test the hardware and software functions of the integrated vehicle, the Flight Operations Control Center, and the ground support crew under launch and flight conditions; and (3) determine the operational characteristics and flight readiness of the vehicle for any subsequent flight tests. Additional operational objectives included turning the vehicle around and flying it again after only seventy-two hours and performing the rotational maneuver.[65] The three-person Flight Operations Control Center and the fifteen-person ground support crew highlighted the project's "faster, cheaper, smaller" management philosophy. Dan Dumbacher was the NASA program manager for the *DC-XA* project. Previously, he had worked on the Space Shuttle Main Engine at the Marshall Space Flight Center, eventually becoming assistant program manager.[66]

The first *DC-XA* flight took place on May 18, 1996. The vehicle climbed 800 feet (244 m) from the launch stand, then flew laterally for 350 feet (115 m), before landing. The total flight time was about one minute (62 seconds). The flight went as planned until the landing, when the *DC-XA* descended more slowly than expected. The desired landing velocity was around 4 feet (1.2 m) per second. *DC-X* touchdowns varied from 2 feet (0.6 m) per second to as high as 14 feet (43 m) per second, the last caused by the radar altimeter. The problem with slow landings was that the vehicle sat in the backwash from the rocket engines too long, and the base of the vehicle could suffer heat damage. This first *DC-XA* landing started a small fire on the exterior of the vehicle, which the crew promptly extinguished. One of the vehicle's four body-flaps (hinged square control surfaces, one on each side of the conical vehicle near its base) suffered damage and had to be replaced.

Following repairs, the *DC-XA,* now dubbed the *Clipper Graham* in honor of Daniel Graham, who had passed away in 1995, undertook its second flight on June 7, 1996. The craft climbed vertically to an altitude of 1,935 feet (590 m), then flew laterally for 590 feet (180 m) before descending tail first onto the desert floor. Total flight time was 63.6 seconds. The flight tested the differential global positioning system (DGPS) that provided positional data to the *Clipper Graham*'s navigational system. Signals from GPS satellites and a ground station provided data to determine precisely the position of the *DC-XA.* The following day, June 8, 1996, the *Clipper Graham* flew again, only twenty-six hours after the previous flight. This was a record turnaround time for any rocket-powered vehicle. The third flight also set a new altitude and duration record. The vehicle reached an altitude of 10,302 feet (3,140 m) and flew for 142 seconds.

The fourth flight took place the following month, on July 31, 1996. The vehicle reached an altitude of 4,100 feet (1,250 m) during a flight that lasted 140 seconds. The *Clipper Graham* completed its planned flight profile, which included an arc-like sweeping maneuver from a nearly upright position, before descending base first. All of the vehicle's components functioned normally. However, 98 seconds into the flight, at about 400 feet (122 m) in the air, the *DC-XA* computer commanded deployment of its landing gear. Over the next 5 seconds, three of the four legs successfully deployed. First landing gears 1 and 4 extended, then landing gear 3 deployed one full second later. Landing gear 2 failed to operate. Nonetheless, the craft landed safely on three of its four legs. When the vehicle detected weight on its landing gear, a signal that it had landed, the *Clipper Graham* turned off its engines, then toppled over toward the position of landing gear 2.

Upon impact, a series of three explosions spaced over 90 seconds destroyed the vehicle. The first explosion ignited the composite shell and the avionics rack. Ten seconds later, the fire suppression system began releasing water. Then a second explosion of liquid oxygen from the aluminum-lithium tank rocked the mishap scene. The fire suppression system shut down when it ran out of water, but the fire continued to burn. About one minute after the second explosion, the hydrogen tank blew up and scattered aeroshell and hydrogen tank composite material over the accident scene.

The explosions and fires totally destroyed the *Clipper Graham.* The upper two-thirds of the composite aeroshell, the aluminum-lithium liquid oxygen tank, the composite liquid hydrogen tank, the composite intertank, avionics, nose cone, and parachute recovery systems were in ruins. The lower third of the vehicle aeroshell, containing the four RL-10 engines and the auxiliary propulsion system, was charred and covered with soot. The engines and auxiliary propulsion system were the only items apparently recoverable. Videotapes of the flight and still photographs of the wreckage showed that landing gear 2 had failed to deploy. Also, post-mishap inspection found the landing gear to be stowed and the pneumatic brake line not connected.[67] NASA established a mishap investigation board to look into the causes of the incident. Whatever their conclusions, the fourth flight marked the end of the *Clipper Graham.*

The NASA board, headed by former astronaut Vance D. Brand of NASA's Dryden Flight Research Center, concluded that the primary cause of the accident was that a technician had not connected the brake line on the helium pneumatic system for landing gear 2. The unconnected brake line prevented the brake mechanism from becoming pressurized and releasing the brake. The result was that landing gear 2 did not extend, the vehicle became unstable when it landed, and the *DC-XA* toppled onto its side, exploded, and burned.

A number of causes contributed to the failure, according to the mishap investigation board. The main cause was human error and the design of the system for stowing the gear. Technicians verified the integrity of the helium brake line, then had to disconnect and reconnect the line. However, they performed no subsequent check to verify the integrity of the system. Landing gear stowage was never identified as a critical process, and no special steps were taken to ensure the readiness of that system for flight. Furthermore, the board found that, during the gear stowage process, technicians made no record of checking off steps. There was no evidence of another technician cross-checking that work, either. Indeed, distraction or interruption of the technician during gear stowage operations likely contributed to the brake line not being connected.[68]

In short, the mishap investigation board largely blamed human error for the demise of the *Clipper Graham.* The failure to connect the landing gear brake line was not an isolated incident. While preparing the *DC-XA* for its third flight, a hydraulics technician had discovered the same failure to connect the landing gear brake line. The technician brought the matter to the attention of the other technicians, and they corrected the problem before flight. However, nobody bothered to document the incident, and nobody changed the procedure to highlight the issue. "This appears," the NASA board concluded, "to have been a missed opportunity to tighten up the controls of this critical process."[69] Ironically, a program built around achieving operational goals had failed in the very area of operations.

The board also found fault with the project's management approach. They pointed out that because the *DC-XA* design and operational procedures had been "driven by rapid development and low cost," a number of imprudent practices resulted. "Accordingly, a minimum number of personnel were involved in operations. Also, design was single string, and there was just one flight test vehicle. There was strong reliance on good people but not a lot of margin for human error afforded by the vehicle preparation process. The McDonnell Douglas Rapid Prototyping Guidelines or implementation thereof for the *DC-XA* may have gone too far in the direction of sacrificing quality and reliability. This rapid prototyping concept should be revisited from an operations perspective."[70]

In other words, "faster, cheaper, smaller" had failed to control "human error." This was the first of many reports that assailed Goldin's "faster, better, cheaper" mantra. A mishap investigation board instituted following the loss of the *Mars Polar Lander* and the *Mars Climate Orbiter* criticized "faster, better, cheaper" in passing, and Jet Propulsion Laboratory project manager Anthony Spear, who had headed the successful *Pathfinder* mission, led an investigation of "faster, better, cheaper" itself.[71]

DC-X program manager Jess Sponable saw the failure of the *DC-XA* in a different light. He recalled the "demoralized" McDonnell Douglas test team that spent weeks working twelve-hour days at 110°F "only to watch NASA give the X-34 contract to OSC [Orbital Sciences Corporation] and the X-33 contract to Lockheed Martin." These were contracts that McDonnell Douglas should have won because of its *DC-X* experience. Orbital and Lockheed had "far less defined designs [than the *DC-X*] that would cost more and were far more risky." "With a demoralized MD [McDonnell Douglas] team in the desert, it was only a matter of time before someone made a critical mistake . . . and sooner than expected."[72]

Many fans of the *DC-X* and the *DC-XA* had hoped to see the vehicle on exhibit at the Smithsonian Institution one day.[73] Others wanted to fly it until, well, it crashed. The *Clipper Graham*'s demise ended all such discussion. Now the friends of the vehicle and the vision for which it stood wanted NASA to build another one and to keep flying. NASA funding remained limited, though. The agency decided that there was just not enough money to build a replacement craft, the price tag for which was probably $120 million.[74] Instead, it would spend nearly a billion dollars on the *X-33* single-stage-to-orbit advanced technology demonstrator. Like the *DC-XA,* the *X-33* program highlighted NASA's failure to inculcate "faster, cheaper, smaller" into its culture and its lack of institutional commitment to realizing single-stage-to-orbit transport.

Conclusion

The *Clipper Graham* that fell over and burned on a New Mexico launch pad represented the final stage in the evolution of a vision. The vision, conceived as early as the nineteenth century, remained idealistic and naïve until the cold war brought funding, military need, and the necessary technological development. The possibility of dramatic reductions in rocket operational costs enhanced the practicality and desirability of the dream. The goal of operating and maintaining rockets like aircraft was the beginning of a revolution in thinking about how one should design and operate launch systems. The *DC-X* became the standard bearer of that revolution. It continued into the era that followed the cold war because it successfully shed its cold war identity and acquired a new one more consonant with the postwar world.

If ever there was a time for NASA to undertake the development of a single-stage-to-orbit vehicle to replace the shuttle, it was during the Clinton administration. NASA Administrator Dan Goldin became a champion of single-stage-to-orbit transport and favored the "faster, cheaper, smaller" managerial structure of the SDIO and the *DC-X* program in particular. One can trace the disappointing and frustrating outcome of NASA's interest in single-stage-to-orbit transport to limited funding and the choice of contractor for the *X-33* project. The decision to award the contract for Phase II to Lockheed Martin appeared to make no sense, considering the successful *DC-X* and *DC-XA* experience of McDonnell Douglas.

About a hundred people from industry, the air force, NASA, the Department of Commerce, the Federal Aviation Administration, and various universities made up the seven *X-33* evaluation teams. However, the selection of the winner was entirely up to Gary Payton.[1] "Because of the greater technical content planned for Phase II, the greater traceability of the technologies to be demonstrated by its X-33 to its RLV [reusable launch vehicle], and the thoroughly developed decision process for entering a Phase III which achieves the objects of the RLV program," Payton wrote, "I selected Lockheed Martin to receive the Cooperative Agreement for Phase II of the X-33 Program."[2]

The selection was surprising. While McDonnell Douglas had been working on the *DC-X* since August 1990, Lockheed had only David Urie's sketch of a lifting body vehicle with an aerospike engine in July 1992. Over the next two years, Urie and a handful of colleagues working full and part time adopted the body shape of the *X-24A* and tested it in the company's wind tunnel, using only internal Lockheed funds.[3] The weaknesses of the Lockheed *X-33* proposal quickly became evident when the Skunk Works began building the *X-33* in 1996.

On top of control and stability issues, the lack of critical wind tunnel data, and severe weight growth, the project faced a shortage of adequate systems engineering and communication barriers resulting from the Skunk Works' unfamiliarity with collocating projects with subcontractors dispersed about the country. Later the weaknesses recognized by the NASA evaluation teams surfaced, namely the technological difficulties with the aerospike engine and the geometrically complex composite liquid hydrogen tanks that failed on several occasions. In March 2001, NASA terminated the *X-33* project, after Lockheed Martin refused to invest any more of its increasingly scarce funds.[4] On January 7, 2002, workers in Palmdale, California, began disassembling the *X-33*, although NASA and the Lockheed Martin industry team had not agreed yet on the distribution of the vehicle's components.[5]

With the end of the *X-33*, NASA's efforts to develop a single-stage-to-orbit vehicle came to a halt. The Space Launch Initiative (SLI), the latest in a long series of shuttle replacement schemes, planned to fund only two-stage-to-orbit concepts. The SLI was a return to NASA's traditional way of conducting projects. It asked to spend $4.5 billion in the first five years of a ten-year program and dispersed funding to dozens of small entrepreneurial firms as well as to the usual large aerospace corporations.[6] SLI symbolically acknowledged NASA's failure to inculcate "faster, better, cheaper" into its institutional culture. Meanwhile, the Defense Department and industry (Boeing and Lockheed Martin) were developing the *Delta IV* and *Atlas V* throwaway rockets in the Evolved Expendable Launch Vehicle program.

The elections of November 2000 appeared to herald changes in the civilian and military space programs established during the Clinton years. After Goldin retained his position during a long transition period that hinted at Bush's disinterest in space, Sean O'Keefe became the new NASA administrator on December 21, 2001. O'Keefe had been the deputy director of the Office of Management and Budget since March. Before that, he had been secretary of the navy, Pentagon comptroller and chief financial officer, and a member of the Senate Committee on Appropriations staff.

Within a month of assuming charge of the space agency, O'Keefe embarked on a series of changes that hinted at a possible new direction in space policy. Clinton's national space launch policy had assigned the development of reusable launchers to NASA and expendable rockets to the Defense Department. Without instituting any new policy, in January 2002, O'Keefe's NASA and the air force initiated a joint study on the possibility of developing new reusable launch vehicles. At an industry briefing and in other venues, O'Keefe made it clear that he wanted to increase cooperation between the Defense Department and NASA in technology development.[7]

While O'Keefe was drafting NASA once again into military service, Defense Secretary Donald Rumsfeld announced the revival of President Reagan's space-based missile defense system. Moreover, in recognition of the national priority that Bush gave to missile defense, Rumsfeld announced the elevation of the Ballistic Missile Defense Organization to agency status and its new designation, the Missile Defense Agency (MDA), on January 4, 2002.[8] Bush publicly criticized the limits of the 1972 ABM Treaty and went beyond President Reagan when he announced in December 2001 that the country was withdrawing from the treaty, a goal of the conservative space agenda, effectively terminating the treaty on June 13, 2002. Despite these efforts to fulfill the military side of the agenda, Bush remained silent on the commercialization of space. For him, space clearly was not the national priority it had been for the Reagan administration. The result was that single-stage-to-orbit dreams remained but dreams. The Quixotes still had a quest, but no steed, as the Sancho Panzas laid claim to the jousting arena. Expendable launchers and the shuttle continued their uninterrupted reign.

Although the quest for single-stage-to-orbit transport is at heart a technological one, technological progress has been less decisive of its failure or success than has been the political arena. Macropolitical events shaped and determined the SDIO's SSTO program from the start. These included the end of the cold war, the revision of the Strategic Defense Initiative, the subsequent abandonment of Brilliant Pebbles, and the Clinton administration's launch policy. At the micropolitical level, fighting in Congress over budget cuts and competing launch systems, as well as the maneuverings of congressional staff members and Pentagon bureaucrats, threatened the flights of the *DC-X* and ultimately abolished Phase III.[9]

The cold war and its conclusion were vital threads in our story's fabric. The cold war made unprecedented demands for technology, project management, and funding. No expense was to be spared in protecting America from any conceivable threat in the global struggle. As a result, those developing military space technologies were less cost conscious than their commercial counterparts because they were oriented toward military missions, not the marketplace. Defending against and surviving the expected nuclear holocaust overrode any possible cost objections. The resulting technology tended to be highly redundant and expensive.

NASA's culture mirrored this same cold war mentality. The technology NASA developed was expensive, and the space agency acquired a culture that emphasized the research, development, design, and construction of large-scale, expensive, and redundant space systems. The Gemini, Mercury, and Apollo programs, the design and construction of the shuttle, the space stations *(Freedom* and *Alpha),* and the Space Exploration Initiative exemplified these systems. They took many years to complete and necessitated the employment of a small "standing army" supervised by several layers of management (often purposely redundant) to fulfill them. This is NASA's "slower, costlier, bigger" style, and it is how the military-industrial complex typically carried out cold war programs. It presumed a national need, the availability of large federal budgets, and a booming economy to keep it going.

With the end of the cold war, the economy softened and budget cuts became the order of the day. Those cuts were a dominant factor in life after the cold war. The fiscal environment encouraged government to adopt "faster, cheaper, smaller" managerial approaches. "Faster, cheaper, smaller" ironically had its roots in the military-industrial complex and the cold war. Furthermore, the SDIO did not invent it. One of the earliest organizations to establish a "faster, cheaper, smaller" managerial style was MIT's Lincoln Laboratory, created in 1951. Its management structure consisted only of the director, the division heads, and the group leaders. It placed project responsibility and authority with the group leaders, as well as the tasks of strategic planning, seeking project funding, and recruiting new staff. Project teams pulled members from diverse groups and divisions on an ad hoc basis, and disbanded at the project's conclusion.[10]

Goldin's application of "faster, cheaper, smaller" largely failed. Work on both the *X-33* and the *X-34,* a two-stage experimental reusable launcher, ended prematurely, with neither vehicle flown or even completed. Goldin's restructuring and shrinking of the Space Exploration Initiative into a series of robotic missions to Mars also ended in embarrassing failures. These failures were ironic. One of the advantages of "faster, cheaper, smaller" that Goldin touted was that, by lowering the cost of building spacecraft, the agency could build multiple spacecraft. If one spacecraft failed, less money was lost, and one could afford a replacement spacecraft. But NASA never took advantage of the redundancy advantage of "faster, cheaper, smaller." On the other hand, redundancy had been a key factor in inflating costs on "slower, costlier, bigger" projects, but it allowed missions to succeed, even when key hardware elements failed.

An excellent example of a "faster, cheaper, smaller" program is the one at the center of this narrative, the SDIO's SSTO program. Its statement of work placed it squarely within the context of the cold war and the Strategic Defense Initiative, specifically Brilliant Pebbles. However, when the *DC-X* rolled out on April 3, 1993, the world had changed. The cold war was over. The USSR no longer existed. Bill Clinton was president. What would be the role of this cold war program in the New World Order? Or better, to paraphrase a spy novel of the 1960s, how did the *DC-X* manage to come in out of the cold war?

One obvious and simple answer is economics. The *DC-X* never cost more than $35 million per year (usually much less), while the NASP annual budget measured in the hundreds of millions. One could argue that post-bellum budget cuts favored small programs over large ones. This explanation, however, is far too simplistic. Neither Congress nor American politics works that neatly. Although the *DC-X* program was fiscally small, it still came under attack. After winning its fight for funding in Congress, the program found the Pentagon bureaucracy withholding its funds in favor of another launcher program. Program foes later switched to arguments of technological unfeasibility to block funding.

The *DC-X* survived in part because it had a support network that consisted of congressional representatives, senators, and staff members, government and industry program officers, and the space movement. In Congress, Dana Rohrabacher and the New Mexico delegation (especially Sen. Pete Domenici) lent their votes and leverage. Bob Walker and Newt Gingrich joined the fight to save the program. Tim Kyger, on Rohrabacher's staff, was the central figure who coordinated efforts and plotted strategies. The SDIO program managers, Pat Ladner and Jess Sponable, and Bill Gaubatz, from McDonnell Douglas, briefed senators, representatives, staff members, and anyone else deemed vital to the cause.

Perhaps the most important member of the space movement to help the program was Daniel Graham. Renaming the NASA vehicle the *Clipper Graham* attested to his role in starting and promoting the project. He played a critical role by arranging the meeting with Vice President Dan Quayle. He then sustained support through his High Frontier organization, its newsletter, and the Space Transportation Association, as well as through his personal correspondence with key political and governmental figures. In addition, Jerry Pournelle, the head of the Citizens' Advisory Council, maintained an e-mail group that allowed supporters to discuss issues among themselves and to build a consensus on what steps to follow in defending the program.

Other space groups that helped to keep the *DC-X* project alive included the L-5 Society, the National Space Society (and its lobbying arm SpaceCause), the Aerospace States Association, and especially the Space Frontier Foundation and Henry Vanderbilt's Space Access Society. These organizations offered their newsletters, telephone trees, fax machines, manpower, and other resources in support of the ongoing struggle. It is important to recall that the space movement, not McDonnell Douglas or the SDIO, financed the campaign to defend the *DC-X*.

It is easy to see this support network—bureaucrats, corporate officials, representatives, senators, their staff members, and lobbyist organizations (the space movement)—as typical of how our government works. What made this network atypical was that one could best describe the program supporters not as the *Clipper* lobby, but as the friends and fans of the *DC-X*. The words "friends" and "fans" more than "supporters" convey the emotional attachment people had to the vehicle, to the program, and to the feasibility of single-stage-to-orbit transport. They did not experience this emotional attachment because the project started within Pournelle's group, or because it had strong political connections to the Reagan administration and the military-industrial complex. Certainly, officers and members of the Space Frontier Foundation and the Space Access Society did not think that they were fighting to save the military-industrial complex or to further a political agenda of the Right, even though the SDIO unequivocally was a vital part of the military-industrial complex and a creation of the Reagan administration.

The *DC-X* touched the imaginations of both those who worked on the project, whether in Congress or at the SDIO or at McDonnell Douglas, and those who supported the effort as either officers or members of various space groups. An exuberant enthusiasm expressed itself as voluntary effort in newsletters, telephone trees, e-mail groups, faxes, and other resources. *DC-X* fandom expressed itself in sweat bands, T-shirts, the *Clipper* sipper water bottle, refrigerator magnets, decals, patches, mouse pads, and a host of other memorabilia. Although public affairs offices regularly distribute refrigerator magnets, decals, and the like, as NASA did on the *X-33* program, for instance, they do not reverberate with the "team" spirit that the friends and fans of the *DC-X* experienced.

Space group members supported the *DC-X* and single-stage-to-orbit transport because of the promise of greatly lower launch costs. Getting into space was and is what space group members want to do. They want to see space explored, opened for commercial exploitation, and eventually colonized. They are frustrated by our failure to return to the Moon, and they want to see humans walking on Mars as soon as possible (which means tomorrow, for the most fervent). With economical launch costs, space will become truly the "final frontier" for them. It is a frontier reminiscent of the American frontier, complete with asteroid miners. These also are the space visions that science fiction authors conjure. As sociologist William Sims Bainbridge succinctly wrote in 1976: "Science fiction is the popular culture of the Spaceflight Movement."[11] Not surprisingly, then, scientists and science fiction writers and their audience constitute the great majority of space interest group membership. Science fiction writers for their part create entertaining stories based on science fact and fantasy that feed the imaginations of their fans inside and outside the space movement, and writers actively participate in the space movement as members and officers.

While science fiction and an enthusiasm for space provided the friends and fans of the *DC-X*, they alone did not rescue the project. Arguments about lowering launch costs so that private enterprise can mine asteroids do not win votes. Rather, the *DC-X* survived the end of the cold war by exchanging its cold war identity for a new one that dovetailed with postwar politics and realities. The Clinton administration's National Space Transportation Policy stipulated: "The National Aeronautics and Space Administration will be the lead agency for technology development and demonstration for next generation reusable space transportation systems, such as the single-stage-to-orbit concept." While this and the National Performance Review formed the policy framework for transferring the project to NASA, the agency itself had been readied for it through Goldin's adoption of "faster, cheaper, smaller" and the results of the Access to Space study. By transforming the role of NASA, Clinton administration policy and NASA provided a civilian institutional home for the project and thus facilitated its safekeeping in the fiscally conservative post–cold war era.

This history also has been about the formulation and implementation of the conservative space agenda within the country's broader political turn to the Right. That agenda was ideological, not partisan. As Roger Launius and Howard McCurdy have pointed out, space exploration enjoyed support from a bipartisan coalition from 1967 into the 1980s. Partisan differences reemerged in the late 1980s and 1990s as part of a larger trend toward partisanship on a range of issues.[12] They also agree on the ideological nature of the space program. "Among all the factors affecting space exploration," Launius and McCurdy wrote, "ideology is the most important. . . . From the beginning of the space age in 1957, the ideological debate over the program has revolved around the expense and direction of the enterprise, particularly the emphasis placed on human spaceflight initiatives as opposed to scientific objectives."[13]

According to Launius and McCurdy, during the 1950s, conservatives endorsed a limited civilian space program focused on scientific research. With the launch of *Sputnik,* liberals clamored for an aggressive space program featuring human spaceflight, sizable federal expenditures, and substantial federal management (NASA). Liberals also pushed the need to garner national prestige, a need eschewed by conservatives, and the need to take on the Soviet Union in a space race. These ideological divisions began to shift as Richard Nixon entered the White House. By approving the shuttle project, Nixon accepted the liberal space agenda of expensive spaceflight; however, he and his cabinet also saw the shuttle's potential for conducting various military missions, which was consistent with the conservative space agenda.[14]

As conservative support shifted toward the space program, Launius and McCurdy explain, liberal support retreated. This "sea change in ideological attitudes toward space . . . drew its strength from the confluence of . . . the changing nature of American liberalism and the conservative embrace of frontier mythology." Kennedy made liberal use of the frontier analogy in his speeches, especially as a rationale for the ambitious Apollo project and the space race with the Soviet Union. Once liberal interest in the cold war waned, so did the necessity of dominating "this new sea." Liberals also increasingly rejected the frontier myth and its implied associations with exploitation and oppression. Conservatives lacked these misgivings about the frontier and embraced the economic benefits and material progress associated with the frontier myth.[15]

Launius and McCurdy further summarized the ideological shift in these terms:

> The civilian space program began with conservatives embracing a limited undertaking with modest scientific objectives. Liberal Democrats created a crash program supported by a robust aerospace industry. Conservatives found it easier to vote for space spending as the program matured, especially when industrial contracts were directed toward conservative strongholds in the South and West. Liberals did not. By the 1980s, the transformation was complete. Liberals in both parties found themselves opposed to the big space program, while conservatives (increasingly concentrated in the Republican Party) had become the bearers of the Kennedy appeal.[16]

The Reagan years, then, marked the high tide of the conservative adoption of what had been the liberal space agenda under President Kennedy, including the frontier myth and the rhetorical analogy between space and the seas (a "spacefaring nation"). Ronald Reagan, Newt Gingrich, Jim Muncy, and Jerry Pournelle's Citizens' Advisory Council all consciously echoed Kennedy's "by the end of the decade" Apollo challenge, as well as the president's many allusions to the frontier myth and the sea-space analogy. Moreover, for Daniel Graham, Mahan's thoughts on the relationship between defense and commerce were applicable to space. Coincidentally, Mahan's ideas appeared in print as Turner's frontier was closing and the seas (and overseas interests) promised to serve as a new imperial frontier.

In addition to taking up Kennedy's space and frontier rhetoric, promoters of the conservative space agenda embraced the Kennedy era's enthusiasm for large-scale space ventures overseen by NASA. In 1984, for example, Newt Gingrich wrote that he wanted NASA's budget to be maintained at the high Apollo program level, and he endorsed (as Reagan did) both the shuttle and the space station. For his part, Gingrich saw nothing inconsistent with being a conservative and being in favor of such large-scale federal expenditures. A final parallel between the Kennedy space agenda and that of the conservatives of the 1980s was the extension of the cold war into space. President Kennedy hoped to beat the Soviet Union to the Moon in order to achieve a cold war objective. Similarly, President Reagan pursued a cold war victory in space through the Strategic Defense Initiative. Clearly, though, major differences between Kennedy's liberal agenda and the conservative space agenda of the 1980s existed. President Kennedy did not favor business or defense interests over such social issues as poverty, education, and integration, whereas the conservative space agenda promoted business interests, laissez-faire economics, and defense spending over social programs.

Distinguishing between the conservative space agenda and President Clinton's space policy is a frustrating task because the Clinton administration appeared to embrace two aspects of the conservative space agenda: "faster, cheaper, smaller" and the commercialization of space. This stance was a clever political stratagem. The Clinton administration's policy of encouraging space commerce mirrored its broader policy of promoting all business, but especially high-technology business. President Clinton, though, was not implementing the conservative space agenda. Another Democratic president, Jimmy Carter, also had favored business interests. Clinton never strongly supported the other key half of the conservative space agenda, the Strategic Defense Initiative, and his policies never placed business or defense interests above social issues. Thus, by promoting space commerce, Clinton administration policy created the mistaken appearance for many outside the space movement that the conservative space agenda and national space policy were one and the same.

The National Space Transportation Policy placed great importance on expanding the commercial sector's role in space. That policy advanced the legacy of the Reagan administration, which had made the fostering of space commerce a national priority. Indeed, the notion that private enterprise can accomplish undertakings more effectively than government agencies was a key tenet of conservative economic thinking that continued into the 1990s as a theme in debates on such diverse topics as the space station, educational reform, and the future of Comsat. The conservative space agenda and national space policy were indistinguishable. America's turn to the Right (at least in space commercial policy) seemed complete. The first years of the new century witnessed NASA's apparent return to military cooperation, the revival of the space-based missile defense system, and the termination of the 1972 ABM Treaty. The conservative space agenda had triumphed.

Conservative and Clinton administration support for space commerce came at a propitious moment. After floundering in the shadow of the shuttle, the U.S. private launch industry finally came into its own in 1989, as the cold war was coming to a conclusion. In the years immediately following, the commercial launch industry continued to grow in the face of unrelenting competition in the international market. The aerospace industry entered an era of dramatic restructuring and downsizing following the end of the Soviet military threat. The growth in space commerce that took place in the last decade provided the aerospace industry a respite from the impact of post–cold war defense cuts. That growth ultimately may justify building a commercial single-stage-to-orbit vehicle. In the meantime, commercial competition in space has taken the place of national rivalries in the cold war. Russia and China remain U.S. rivals, but in the global launcher market, suggesting that the cold war has transmuted into a "gold" war. In 2000, though, as various high-technology industries began to soften, space commerce, because of its high capital requirements, was one of the first to falter, starting with the failure of Motorola's Iridium constellation.

Meanwhile, too, the supporters of the conservative space agenda grew in number. Significant growth took place among those who advocated, for want of a better term, a libertarian space agenda. Increasingly frustrated by NASA's preference for the corporate dinosaurs of the cold war (such as Lockheed Martin) and the agency's indifference to entrepreneurial start-up firms, the libertarians wanted government to relinquish its control over all aspects of space commerce. Among the individuals mentioned in this story, Gary Hudson (Rotary Rocket) and Rick Tumlinson (President, Space Frontier Foundation) exemplified the libertarian position of opening the space "frontier" for private commercial development without governmental interference or assistance.

The United States has known only the cold war for the past half century. It is only natural that we build with what we have on hand and from what we know. We are still in a transitional stage, emerging slowly from the patterns established in the past. It remains for a future generation capable of seeing with new ideas to break these patterns and to create a true postwar society, unless, of course, the country engages in another protracted struggle.

Appendix: Kelly's Rules

1. The Skunk Works manager must be delegated practically complete control of his program in all aspects. He should report to a division president or higher.

2. Strong but small project offices must be provided both by the military and industry.

3. The number of people having any connection with the project must be restricted in an almost vicious manner. Use a small number of good people (10 percent to 25 percent compared to the so-called normal systems).

4. A very simple drawing and drawing release system with great flexibility for making changes must be provided.

5. There must be a minimum number of reports required, but important work must be recorded thoroughly.

6. There must be a monthly cost review covering not only what has been spent and committed but also projected costs to the conclusion of the program. Don't have the books ninety days late and don't surprise the customer with sudden overruns.

7. The contractor must be delegated and must assume more than normal responsibility to get good vendor bids for subcontract on the project. Commercial bid procedures are very often better than military ones.

8. The inspection system, as currently used by the Skunk Works, which has been approved by both the Air Force and the Navy, meets the intent of existing military requirements and should be used on new projects. Push more basic inspection responsibility back to subcontractors and vendors. Don't duplicate so much inspection.

9. The contractor must be delegated the authority to test his final product in flight. He can and must test it in the initial stages. If he doesn't, he rapidly loses his competency to design other vehicles.

10. The specifications applying to the hardware must be agreed to in ad-

vance of contracting. The Skunk Works practice of having a specification section stating clearly which important military specification items will not knowingly be complied with and reasons therefore is highly recommended.

11. Funding program must be timely so that the contractor doesn't have to keep running to the bank to support government projects.

12. There must be a mutual trust between the military project organization and the contractor with very close cooperation and liaison on a day-to-day basis. This cuts down misunderstanding and correspondence to an absolute minimum.

13. Access by outsiders to the project and its personnel must be strictly controlled by appropriate security measures.

14. Because only a few people will be used in engineering and most other areas, ways must be provided to reward good performance by pay not based on the number of personnel supervised.

Notes

Introduction

1. Winged vehicles also have a strong appeal to NASA engineers and managers.

2. I have borrowed the notion "true spaceship" from Max Hunter, in particular from his manuscript "The SSX: A True Spaceship," October 4, 1989, file 204, *X-33* Archive (see my bibliographic essay).

3. Richard P. Hallion, "ASSET: Pioneer of Lifting Reentry," in Hallion, ed., *The Hypersonic Revolution: Eight Case Studies in the History of Hypersonic Technology*, vol. 1, *From Max Valier to Project Prime, 1924-1967* (Dayton: Special Staff Office, Aeronautical Systems Division, Wright-Patterson AFB, 1987), pp. 450, 454, 459, 483.

4. Roger E. Bilstein, *The American Aerospace Industry: From Workshop to Global Enterprise* (New York: Twayne Publishers, 1996), p. 84.

ONE: The Reagan Revolution

1. Ronald Reagan, "State of the Union Message, January 25, 1984," *Public Papers of the Presidents of the United States: Ronald Reagan, 1984* (Washington, DC: Government Printing Office, 1985), p. 90.

2. For my discussion of these political events, I have relied on William C. Berman, *America's Right Turn*, 2d ed. (Baltimore: Johns Hopkins University Press, 1998), pp. 2-3, 6-8. Also useful were Mary C. Brennan, *Turning Right in the Sixties: The Conservative Capture of the GOP* (Chapel Hill: University of North Carolina Press, 1995); Dan T. Carter, *The Politics of Rage: George Wallace, the Origins of the New Conservatism, and the Transformation of American Politics* (New York: Simon and Schuster, 1995); idem, *From George Wallace to Newt Gingrich: Race in the Conservative Counterrevolution* (Baton Rouge: Louisiana State University Press, 1997); and Godfrey Hodgson, *The World Turned Right Side Up: A History of the Conservative Ascendancy in America* (Boston: Houghton Mifflin, 1996).

3. Berman, pp. 21, 39. For U.S. political and social history during the 1970s, see Peter Carroll, *It Seemed Like Nothing Happened: America in the 1970s* (New Brunswick: Rutgers University Press, 1990), while Douglas Hibbs, *The American Political Economy: Macroeconomics and Electoral Politics* (Cambridge: Harvard University Press, 1987), presents well-researched analysis of inflation's impact on presidential politics during the presidencies of Ford, Carter, and Reagan. I have discussed the invasion of the U.S. industrial gas market by European firms in Butrica, *Out of Thin Air: A History of Air Products and Chemicals, Inc., 1940-1990* (New York: Praeger, 1990), pp. 244-247.

4. On Paul Volcker's role in undermining Carter's hopes in 1980, see William Grieder, *Secrets of the Temple: How the Federal Reserve Runs the Country* (New York: Simon & Schuster, 1987).

5. Allen M. Kaufman, Marvin J. Karson, and Jeffrey Sohl, "Business Fragmentation and Solidarity: An Analysis of PAC Donations in the 1980 and 1982 Elections," in Alfred A. Marcus, Allen M. Kaufman, and David R. Beam, eds., *Business Strategy and Public Policy* (New York: Quorum Press, 1987), pp. 119-136.

6. Kevin Phillips documents the dramatic transfer of wealth upward in the 1980s in *The Politics of Rich and Poor: Wealth and the American Electorate in the Reagan Aftermath* (New York: Random House, 1991).

7. Norman R. Augustine, "Space Commercialization and Industry," p. 4 in Ted W. Jensen, gen. ed., *Space, The Next Ten Years* (Colorado Springs, CO: United States Space Foundation, 1985).

8. Paul Neurath, *From Malthus to the Club of Rome and Back: Problems of Limits to Growth, Population Control, and Migrations* (Armonk, NY: M. E. Sharpe, 1994) reviews the limits-to-growth "debate" from the eighteenth century to the present, including the Club of Rome, and has a bibliography of the literature. Robert McCutcheon, *Limits to a Modern World: A Study of the Limits to Growth Debate* (London: Butterworths, 1979) provides a contemporary overview of the "debate."

9. James Muncy, interview by author, tape and transcript, January 12, 1999, Washington, DC, p. 66, NASA Historical Reference Collection (hereafter NHRC).

10. Newt Gingrich and James A. M. Muncy, "Space: The New Frontier," p. 61, in Paul M. Weyrich and Connaught Marshner, eds., *Future 21: Directions for America in the 21st Century* (Greenwich, CT: Devin-Adair, 1984).

11. Ibid., p. 62; Newt Gingrich, *Window of Opportunity: A Blueprint for the Future* (New York: Tom Doherty Associates, Inc., 1984), p. ix; Muncy, interview, p. 67.

12. Gingrich, pp. 1, 7-9.

13. Gingrich, pp. 10, 27, 46, 49-50, 65-66.

14. Gingrich, p. 52.

15. Gingrich, pp. 53-54; Gingrich and Muncy, p. 62.

16. Gingrich, pp. 59-60, 52.

17. Gingrich, p. v.

18. Muncy, interview, p. 6.

19. Jerry Pournelle and Stefan Possony, *The Strategy of Technology: Winning the Decisive War* (Cambridge, MA: University Press of Cambridge, 1970).

20. Jerry Pournelle, "Draft Response," December 16, 1998, pp. 1, 2, 6, file 754, *X-33* Archive (see my bibliographic essay).

21. Gingrich and Muncy, p. 69; Gingrich, p. 269; and Muncy, interview, p. 7.

22. Cited in Gingrich, p. 11. See Herman Kahn, *World Economic Development: 1979 and Beyond* (Boulder, CO: The Westview Press, 1979): Kahn, William Brown, and Leon Martel, *The Next 200 Years: A Scenario for America and the World* (New York: Morrow, 1976): and Kahn, *The Coming Boom: Economic, Political, and Social* (New York: Simon and Schuster, 1982). G. Harry Stine also contributed to this futurist literature in his *Third Industrial Revolution* (New York: Putnam, 1975).

23. Cited in Gingrich, p. 2. Alvin Toffler, *Future Shock* (New York: Random House, 1970), and Toffler, *The Third Wave* (New York: Morrow, 1980).

24. Gingrich, "Window of Opportunity," *Futurist* 19, no. 3 (June 1985): 9-15.

25. Muncy, interview, pp. 3-5.

26. Erik K. Pratt, *Selling Strategic Defense: Interests, Ideologies, and the Arms Race* (Boulder: Lynne Rienner Publishers, 1990), p. 96.

27. John F. Kennedy, address at Rice University, September 24, 1962, *Public Papers of the Presidents: John F. Kennedy, 1962* (Washington, DC: National Archives and Records Service, 1963), p. 329.

28. Pratt, p. 96; Daniel Graham, *High Frontier: A New National Strategy* (Washington, DC: The Heritage Foundation, 1982); Graham, *The Non-Nuclear Defense of Cities: The High Frontier Space-Based Defense Against ICBM Attack* (Cambridge, MA: Abt Books, 1983). On Mahan and his theories, see Robert Seager, *Alfred Thayer Mahan: The Man and His Letters* (Annapolis: Naval Institute Press, 1977) and Alfred Thayer Mahan, *The Influence of Sea Power on History, 1660-1783* (1897; reprint, New York: Dover Publications, 1987).

29. William Edmund Livezey, *Mahan on Sea Power,* rev. ed. (Norman: University of Oklahoma Press, 1980), pp. 42-43.

30. Alfred Thayer Mahan, "Hawaii and Our Future Sea Power," in Mahan, *The Interest of America in Sea Power, Present and Future* (Boston: Little, Brown and Company, 1918), p. 52.

31. Alfred Thayer Mahan, "Possibilities of an Anglo-American Reunion," in Mahan, *The Interest of America in Sea Power,* p. 118.

32. Livezey, pp. 45, 48, 52.

33. Seager, p. 121.

34. Mahan, "Hawaii and Our Future Sea Power," p. 52.

35. W. D. Kay, "Space Policy Redefined (Again)," in "Contested Ground: The Historical Debate over NASA's Mission" (manuscript provided by author), pp. 28, 31.

36. "National Space Policy," July 4, 1982, file 386, *X-33* Archive.

37. Kay, "Space Policy Redefined (Again)," pp. 32, 35.

38. Christopher Simpson, *National Security Directives of the Reagan and Bush Administrations: The Declassified History of U.S. Political and Military Policy, 1981-1991* (Boulder, CO: Westview Press, 1995), pp. 136-143 (classified version) and 144-150 (unclassified version); Kay, "Space Policy Redefined (Again)," pp. 32, 35.

39. Lyn Ragsdale, "Politics Not Science: The U.S. Space Program in the Reagan and Bush Years," in Roger D. Launius and Howard E. McCurdy, eds., *Spaceflight and the Myth of Presidential Leadership* (Urbana: University of Illinois Press, 1997), p. 133.

40. Judy A. Rumerman and Stephen J. Garber, *Chronology of Space Shuttle Flights, 1981-2000,* HHR-70 (Washington, DC: NASA History Division, Office of Policy and Plans, NASA Headquarters, October 2000), p. 5.

41. Note to Mike Deaver, May 19, 1982, file 696, *X-33* Archive.

42. "Space Launch Policy Working Group Report on Commercialization of U.S. Expendable Launch Vehicles," April 13, 1983, pp. 1-2, 34, file 696, *X-33* Archive.

43. Rosalind A. Knapp to David A. Stockman, December 12, 1983, file 696, *X-33* Archive.

44. Draft National Security Decision Directive, April 22, 1983, file 696, *X-33* Archive.

45. Craig L. Fuller to Richard G. Darman et al., note and attachment, "Space Commercialization," August 2, 1983, and agenda, space commercialization meeting, August 3, 1983, file 696, *X-33* Archive.

46. "National Policy on the Commercial Use of Space," July 20, 1994, file 386, *X-33* Archive.

47. Mark Chartrand, "On the New Conestoga Trail—A Personal Account," *Space World* S-11-227 (November 1982): 5; Kim G. Yeltson, "Evolution, Organization, and Implementation of the Commercial Space Launch Act and Amendments of 1988," *The Journal of Law & Technology* 4 (1989):119.

48. "Space Launch Policy Working Group Report on Commercialization of U.S. Expendable Launch Vehicles," April 13, 1983, pp. 30-31, file 696, *X-33* Archive.

49. Ibid., p. 6.

50. "Executive Order: Commercial Expendable Launch Vehicle Activities," attached to Michael J. Horowitz to Robert Kimmitt, December 12, 1983, and Rosalind A. Knapp to David A. Stockman, December 12, 1983, file 696, *X-33* Archive.

51. Robert C. McFarlane to Craig Fuller, "Report and Proposals from the Commercial Space Group," memorandum, December 23, 1983, file 696, *X-33* Archive.

52. U.S. House of Representatives, *Commercial Space Launch Act,* 98th Cong., 2d sess., Report 98-816 (Washington, DC: Government Printing Office, 1984), p. 9.

53. Copies of H.R. 3942 and S.560 in file 388, *X-33* Archive.

TWO: Commerce on the High Frontier

1. W. D. Kay, "Space Policy Redefined (Again)," in "Contested Ground: The Historical Debate over NASA's Mission," (manuscript provided by author), p. 11.

2. Pamela E. Mack, "LANDSAT and the Rise of Earth Resources Monitoring," in Mack, ed., *From Engineering Science to Big Science: The NACA and NASA Collier Trophy Research Project Winners,* NASA SP-4219 (Washington, DC: NASA, 1998), pp. 237, 242.

3. Ibid., pp. 248-249.

4. Cabinet Council on Commerce and Trade to the President, "Transfer of the Civil Space Remote Sensing Systems to the Private Sector," memorandum, February 28, 1983, file 387, *X-33* Archive (see my bibliographic essay).

5. Mack, p. 250.

6. David J. Whalen, "Billion Dollar Technology: A Short Historical Overview of the Origins of Communications Satellite Technology, 1945-1965," pp. 106, 111 in Andrew J.Butrica, ed., *Beyond the Ionosphere: Fifty Years of Satellite Communication,* NASA SP-4217 (Washington, DC: NASA, 1997).

7. "NASA Commercial Space Policy," October 1984, "Introduction," file 386, *X-33* Archive; John R. Carruthers, briefing charts, "Innovation of Space Technology Through Joint Endeavors between NASA and Private Industry," March 28, 1980, American Astronautical Society 18th Goddard Memorial Symposium, file 383, *X-33* Archive; "NASA Guidelines Regarding Early Usage of Space for Industrial Purposes," June 25, 1979, file 386, X-33 Archive.

8. "NASA Commercial Space Policy," October 1984, pp. ii, v, and "Summary of Policy Initiatives" and "Research and Development Initiatives," file 386, *X-33* Archive.

9. Ibid., "Summary of Policy Initiatives"; Dexter C. Hutchins, "Entrepreneurs Aim for Outer Space," *Venture* n.v. (September 1980): 48, file 383, *X-33* Archive; and James Drummond, "Private Enterprise's Big Bet: Space Ventures Will Pay Off," *Houston Chronicle,* December 25, 1983, sec. 4, p. 1, file 383, *X-33* Archive.

10. NASA Special Announcement, "Establishment of the Office of Commercial Programs," September 11, 1984; NASA News Press Release 87-126, "Assistant Administrator Gillam to Retire from NASA," August 19, 1987; "NASA Commercial Space Policy," October 1984, "Summary of Policy Initiatives;" and Isaac T. Gillam IV, "Encouraging the Commercial Use of Space and NASA's Office of Commercial Programs," *NASA Tech Briefs,* n.v. (Spring 1985): 14-15, all in file 383, *X-33* Archive; John M. Cassanto, "CCDS Shock Waves," *Space News,* January 24-30, 1994, p. 21.

11. Ronald Reagan, "State of the Union Message, January 25, 1984," *Public Papers of the Presidents of the United States: Ronald Reagan, 1984* (Washington, DC: Government Printing Office, 1985), p. 90. Excerpts also in file 386, *X-33* Archive.

12. Kay, "Space Policy Redefined (Again)," pp. 64-67; Howard E. McCurdy, *The Space Station Decision: Incremental Politics and Technological Choice* (Baltimore: Johns Hopkins University Press, 1990), pp. 179-180.

13. Mitch Lobrovich, "Trying to Help People Make a Buck in Space," *The Dallas Morning News,* July [date unknown], 1981, p. 63, photocopy, file 383, *X-33* Archive; "NASA Consultants Report on Space Station Customer Prospects," *Space Business News,* June 18, 1984, pp. 4-5, photocopy, file 383, *X-33* Archive; Adam Gruen, Chapter 3, "Within a Decade," in "The Port Unknown: A History of Space Station *Freedom,*" (unpublished manuscript provided by author, 1991), pp. 14-15, NASA Historical Reference Collection (hereafter NHRC). See also "NASA Guidelines for United States Commercial Enterprises for Space Station Development and Operations," August 7, 1986, file 383, *X-33* Archive.

14. Andrew J. Butrica, *To See the Unseen: A History of Planetary Radar Astronomy,* NASA SP-4218 (Washington, DC: NASA, 1996), pp. 24-25.

15. Butrica, *Beyond the Ionosphere,* p. xv.

16."Orbital Sciences Offers Upper Stages," *Aviation Week & Space Technology* 120, no. 26 (June 25, 1984): 108-115; "Broad Spectrum of Business Involved in Space Commercialization," *Aviation Week & Space Technology* 120, no. 26 (June 25, 1984): 63; Carole A. Shifrin, "Investors Taking Cautious View of Private Programs," *Aviation Week & Space Technology* 120, no. 26 (June 25, 1984): 79, 80; Christopher J. Cohan, Walter B. Olstad, Donald W. Patterson, and Robert Salkeld, *Space Transportation Systems, 1980-2000,* vol. I, AIAA Aerospace Assessment Series (New York: AIAA, 1978), pp. 20-22, file 383, *X-33* Archive.

17. Klaus P. Heiss, "Space: Opportunity and Challenge for Free Enterprise in the Next Decades," October 18, 1979, pp. 19-24, file 383, *X-33* Archive; The Space Transportation Company, "Private Sector Initiatives in Space Commercialization of Expendable Launch Vehicles," 1983, and other SpaceTran materials in file 383, *X-33* Archive.

18. James C. Bennett, "The Second Space Race," *Reason* n.v. (November 1981): 27, photocopy, file 383, *X-33* Archive.

19. "Two Firms Ready to Buy, Produce Shuttle Orbiters," *Aviation Week & Space Technology* 120, no. 26 (June 25, 1984): 116-119.

20. John Schaus, Orbital Sciences Corporation, personal communication, April 9, 1999. Schaus was with Fairchild Space at the time. "Fairchild Seeks Agreements on Leasecraft," *Aviation Week & Space Technology* 120, no. 26 (June 25, 1984): 54; Craig Covault, "Lack of Insurance, Customers Halts Fairchild Leasecraft," *Aviation Week & Space Technology* 123, no. 20 (November 11, 1985), pp. 16-17; "Fairchild Expects $60 Million in Subcontracts on Leasecraft," *Defense Daily,* October 30, 1984, p. 311; Memo-

randum for the record, "NASA/Fairchild Leasecraft Joint Endeavor Agreement (JEA) meeting, October 6, 1983," November 1, 1983, file 6,132, NHRC.

21. Michael Feazel, "Sparx Decision Clouds Imaging Projects," *Aviation Week & Space Technology* 120, no. 26 (June 25, 1984): 147-149; "SPARX Seeks ELV Launch After NASA Veto," *Space Business News,* June 18, 1984, pp. 2-3.

22. Mitch Lobrovich, "Putting Business into Outer Space: Will Pioneers Be 'Billionaires of the Future'?" *The Dallas Morning News,* November 21, 1981, sec. H, p. 3; "Broad Spectrum of Business Involved in Space Commercialization," p. 63; "3M Seeks New Materials, Processes," *Aviation Week & Space Technology* 120, no. 26 (June 25, 1984): 65-73; "John Deere Plans Space Iron Research," ibid., pp. 74-77; Edward H. Kolcum, "Company Plans to Manufacture Crystals in Space," ibid., pp. 100-101.

23. John R. Carruthers, "Innovation of Space Technology Through Joint Endeavors between NASA and Private Industry," briefing charts, March 28, 1980, American Astronautical Society 18th Goddard Memorial Symposium, file 383, *X-33* Archive; "Medicine Sales Forecast at $1 Billion," *Aviation Week & Space Technology* 120, no. 26 (June 25, 1984): 52; "More on MDAC Space Test," *Space Business News,* June 18, 1984, p. 7.

24. "Medicine Sales Forecast at $1 Billion," p. 53; Eugene Kozicharow, "Patent Law Finds Space Applications," *Aviation Week & Space Technology* 120, no. 26 (June 25, 1984): 99; "Space Industries, Inc., to Begin Marketing Unmanned Facility," ibid., pp. 116-117; Kay, pp. 242-243; documents on the Industrial Space Facility, file 708, *X-33* Archive.

25. Bennett, pp. 25, 28; Hutchins, p. 50.

26. Richard G. O'Lone, "Starstruck's Problems Spotlight Risks, Opportunities in Space," *Commercial Space* 1, no. 1 (Spring 1985): 60-63; idem, "Bay Area Firms Pursue Booster Designs," *Aviation Week & Space Technology* 120, no. 26 (June 25, 1984): 166-169; "Launch Services," *Space* n.v. (January 1985): 17, file 383, *X-33* Archive. Starstruck became American Rocket Company (AmRoc) before disappearing entirely. Tim Kyger, comments on Andrew J. Butrica, "The Spaceship That Came in from the Cold War" (unpublished manuscript), November 27, 2000, p. 2, file 857, *X-33* Archive.

27. Bennett, pp. 22-23, 31; Hutchins, pp. 48, 50, 52.

28. "Free Enterprise Space Shot," *Time,* June 29, 1981, p. 63, and SSI, "The Percheron Project," June 1981, press information kit, file 89, *X-33* Archive.

29. Gary C. Hudson, Tom A. Brown, David J. Ross, Clifton Horne, and Eric Larsen, "A Modular Low-Cost Launch Vehicle System: The Percheron Project," 6 pp., n.d., file 89, *X-33* Archive; Mark Chartrand, "On the New Conestoga Trail—A Personal Account," *Space World* S-11-227 (November 1982): 4-7; "Rocket Explodes During Test," *The Alliance [Ohio] Review,* August 6, 1981, p. 2, photocopy, file 89, *X-33* Archive.

30. Chartrand, pp. 4-7; Leonard David, "Private Rocketeers to Try Again: A Solid Decision," *Space World* S-2-218 (February 1982): 10; "Conestoga and Ariane: Ups and Downs of the Launch Business," *Science News* 122 (September 18, 1982): 180, file 89, *X-33* Archive; Alton K. Marsh, "Space Services Pushing Conestoga Launch Vehicle," *Aviation Week & Space Technology* 120, no. 26 (June 25, 1984): 163-165.

31. Bennett, p. 24.

32. "Transpace Carriers, Inc." n.d., briefing charts, congressional testimony, and other Transpace Carriers documents, file 384, *X-33* Archive; Paul Kinnucan, "Expendable Launch Vehicles Get a Lift," *High Technology* 4, no. 6 (June 1984): 28-29;

"Transpace Embarks as Venture Capitalized Booster Company," *Aviation Week & Space Technology* 120, no. 26 (June 25, 1984): 140-141.

33. Dennis R. Jenkins, *Space Shuttle: The History of Developing the National Space Transportation System* (Marceline, MO: Walsworth Publishing, 1992), pp. 286-287.

34. Gingrich, *Window of Opportunity: A Blueprint for the Future* (New York: Tom Doherty Associates, Inc., 1984), p. 53.

35. Arianespace, *Arianespace: The World's First Commercial Space Transportation Company* (Paris: Arianespace, 1991); Douglas A. Heyden, Arianespace presentation overheads, n.d.; and Shawn Tully, "Europe Blasts into the Space Business," *Fortune*, May 27, 1985, p. 140, file 177, *X-33* Archive.

36. Tully, p. 140.

37. Ibid., pp. 139-140.

38. Ibid., pp. 139-140; ESA News Release, July 24, 1985, file 14,579, NHRC.

39. General Accounting Office, *NASA Must Reconsider Operations Pricing Policy to Compensate for Cost Growth on the Space Transportation System* (Washington, DC: General Accounting Office, 1982), p. 7.

40. Lyn Ragsdale, "Politics Not Science: The U.S. Space Program in the Reagan and Bush Years," in Roger D. Launius and Howard E. McCurdy, eds., *Spaceflight and the Myth of Presidential Leadership* (Urbana: University of Illinois Press, 1997), p. 149.

41. James M. Beggs to Robert C. McFarlane, July 17, 1984, file 387, *X-33* Archive.

42. Robert C. McFarlane to Elizabeth H. Dole, "STS Pricing Issue," memorandum, June 21, 1984, and Dole to McFarlane, memorandum, June 28, 1984, file 387, *X-33* Archive.

43. James A. Baker III to the Economic Policy Council, "Presidential Policy Directive—Space Commercialization," memorandum, October 6, 1986, file 387, *X-33* Archive; and "Presidential Directive on National Space Policy," fact sheet, February 11, 1988, p. 9, file 386, *X-33* Archive.

44. Kim G. Yelton, "Evolution, Organization, and Implementation of the Commercial Space Launch Act and Amendments of 1988," *The Journal of Law & Technology* 4 (1989): 134.

45. "Presidential Directive on National Space Policy," pp. 7, 9, 10.

46. U.S. Department of Transportation, Office of Commercial Space Transportation, "Annual Report to Congress: Activities Conducted under the Commercial Space Launch Act," 1987, pp. 5-6, file 391, *X-33* Archive; "State of the U.S. Commercial Launch Industry" in Department of Transportation Office of Commercial Space Transportation, *The U.S. Office of Commercial Space Transportation Fifth Annual Report* (Washington, DC: Government Printing Office, 1990), file 393, *X-33* Archive. Hereafter, *Fifth Annual Report.*

47. Stephanie Lee-Miller, "Message from the Director," October 1989, n.p., in *Fifth Annual Report.*

48. "First Commercial Rocket Launch Successful," *Space News Roundup,* March 31, 1989, p. 4, file 10,784, NHRC.

49. NASA Press Release 90-58, April 23, 1990; "Researchers Buoyed by Starfire Mission," *Space News,* May 21-27, 1990, p. 3, photocopy; William Ganoe, "Starfire," *Ad Astra,* March 1989, p. 35; and Action Document Summary, August 29, 1986, file 10,782, NHRC.

50. "State of the U.S. Commercial Launch Industry."

51. Ibid.

52. "SSI Halts Operations," *Space News,* July 9-15, 1990, p. 3, photocopy, and "Private Launch Firm Hits Hard Times," *Washington Post,* July 4, 1990, p. D3, photocopy, file 10,782, NHRC. For background on the financial arrangement between SSI and Houston Industries, Inc., see "Space Services Gets Venture Capital Backing," *Defense Daily,* February 27, 1987, p. 311.

53. "Rocket Away!" *EER Systems Newsletter* 5, no. 1, pp. 1-2, file 10,784, NHRC; "Commercial Launch Only Partial Success for Research Team," *Sun-News* [Las Cruces, NM], September 11, 1992, p. 13, photocopy, file 10,784, NHRC; and "Donald Slayton Dies at 69; Was One of First Astronauts," *The New York Times,* Obituaries, June 14, 1993, n.p., photocopy, file 10,782, NHRC.

54. Jim Hengle, interview by author, tape and transcript, February 25, 1998, Futron Corporation, Bethesda, MD, NHRC, pp. 2, 33; NASA Press Release N95-64, October 12, 1995; Warren Ferster, "Destruct Signal Fails in Conestoga Breakup," *Space News,* October 30, 1995, p. 6, photocopy; "Faulty Internal Signal Cited in Rocket Explosion," [Washington] *Metropolitan Times,* October 30, 1995, p. C4, photocopy; "Debris from Rocket Recovered in Va." *Washington Post,* October 25, 1995, p. D5, photocopy; and "Private Rocket Launch Fails in Explosion after Liftoff," *The Washington Times,* October 24, 1995, p. C5, photocopy, all in file 10,783, NHRC.

55. "State of the U.S. Commercial Launch Industry," n.p., and "International Competition," n.p., in *Fifth Annual Report.*

THREE: Space Warriors

1. *Public Papers of the Presidents of the United States: John F. Kennedy, 1961* (Washington, DC: Government Printing Office, 1962), pp. 403-404. Asif A. Siddiqi, *Challenge to Apollo: The Soviet Union and the Space Race, 1945-1974,* NASA SP-4408 (Washington, DC: NASA, 2000), shows that Soviet military officers soon lost interest in civilian space projects following Sputnik. They felt that civilian projects hurt their attempts to fund military rocketry programs essential to the cold war.

2. Edward Clinton Ezell and Linda Neuman Ezell, *The Partnership: A History of the Apollo-Soyuz Test Project,* NASA SP-4209 (Washington, DC: NASA, 1978), discusses the ASTP in detail.

3. Robert S. Kraemer, *Beyond the Moon: A Golden Age of Planetary Exploration, 1971-1978* (Washington, DC: Smithsonian Institution Press, 2000), makes this point and highlights the role of the Jet Propulsion Laboratory.

4. Paul B. Stares, *The Militarization of Space: U.S. Policy, 1945-1984* (Ithaca, NY: Cornell University Press, 1985), pp. 90, 101, 103.

5. Quoted in Richard P. Hallion and James O. Young, "Space Shuttle: Fulfillment of a Dream," in Hallion, ed., *The Hypersonic Revolution: Eight Case Studies in the History of Hypersonic Technology,* vol. 2, *From Scramjet to the National Aero-Space Plane* (Dayton: Special Staff Office, Aeronautical Systems Division, Wright-Patterson AFB, 1987), p. 1041.

6. Erik K. Pratt, *Selling Strategic Defense: Interests, Ideologies, and the Arms Race* (Boulder, CO: Lynne Rienner Publishers, 1990), p. 21; Donald R. Baucom, *The Origins of SDI, 1944-1983* (Lawrence: University Press of Kansas, 1992), pp. 39, 42-50; John C. Lonnquest and David F. Winkler, *To Defend and Deter: The Legacy of the United States*

Cold War Missile Program, USACERL Special Report 97/01 (Champaign, IL: U.S. Army Construction Engineering Research Laboratories, 1996), pp. 113-115.

7. Pratt, p. 21; Baucom, pp. 54-55, 58-61, 91-92, 96-97.

8. Pratt, pp. 21, 31; Baucom, pp. 61-71.

9. Stares, pp. 107, 109-110, 117-131, 135-136, 145-146.

10. Pratt, p. 53; Baucom, p. 76.

11. Stares, pp. 217-218, 229-239.

12. Clarence A. Robinson, Jr., "Soviets Push for Beam Weapon," *Aviation Week & Space Technology* 106, no. 18 (May 2, 1977): 16-23.

13. J. London and H. Pike, "Fire in the Sky: U.S. Space Laser Development from 1968," IAA-97-IAA.2.3.06, pp. 1-3, paper read at the 48th International Astronautical Congress, October 6-10, 1997, Turin, photocopy, file 40, *X-33* Archive (see my bibliographic essay); Pratt, pp. 16-18; Baucom, pp. 15-17.

14. Pratt, pp. 43, 45, 46; Baucom, pp. 81-82.

15. Clarence A. Robinson, Jr., "Army Pushes New Weapons Effort," *Aviation Week & Space Technology* 109, no. 14 (October 16, 1978): 42-43, 45, 48-49, 51-52.

16. Baucom, p. 118.

17. Baucom, p. 119; Max Hunter, "Strategic Dynamics and Space-Laser Weaponry," October 31, 1977, file 338, *X-33* Archive.

18. Baucom, p. 108.

19. Max Hunter, interview by author, tape and transcript, June 19, 1998, San Carlos, CA: NASA Historical Reference Collection (hereafter NHRC), pp. 73-75; London and Pike, pp. 3-4.

20. Pratt, p. 70; Baucom, pp. 122-124, 126.

21. Pratt, p. 71; Baucom, pp. 126, 127; Hunter, interview, p 80.

22. Pratt, p. 71.

23. Daniel O. Graham, *Confessions of a Cold Warrior: An Autobiography* (Fairfax, VA: Preview Press, 1995), pp. 118-120; Baucom, pp. 145-146, 150.

24. William C. Berman, *America's Right Turn,* 2d ed. (Baltimore: Johns Hopkins University Press, 1998), pp. 67-68; David Vogel, *Fluctuating Fortunes: The Political Power of Business in America* (New York: Basic Books, 1989), pp. 224-225; Dilys M. Hill and Phil Williams, "The Reagan Presidency: Style and Substance," in Dilys M. Hill, Raymond A. Moore, and Phil Williams, eds., *The Reagan Presidency: An Incomplete Revolution?* (New York: St. Martin's Press, 1990), p. 11.

25. Pratt, p. 96; Baucom, p. 164.

26. Graham, *Confessions,* p. 144.

27. Pratt, pp. 96-97.

28. Pratt, pp. 102, 103, 104; Baucom, p. 130.

29. See, for example, Baucom; Graham, *Confessions;* Pratt; Stares; and Edward Reiss, *The Strategic Defense Initiative* (New York: Cambridge University Press, 1992).

30. Pratt, p. 109-110.

31. Baucom, pp. 195-196.

32. Reiss, p. 51.

33. Pratt, p. 111.

34. Gary E. Payton, interview by author, tape and transcript, August 20, 1997, NASA Headquarters, Washington, DC, pp. 19-20, NHRC.

35. Reiss, p. 64.

36. Reiss, pp. 63-64.

37. Quoted in Reiss, p. 65.

38. See, for example, Graham, *Confessions,* p. 148; and Lawrence Feinberg, "Lucasfilm's 'Star Wars' Lawsuit," *The Washington Post,* November 11, 1985, p. B3.

39. Lonnquest and Winkler, pp. 129-130; Reiss, pp. 89-90.

40. Reiss, p. 88.

FOUR: *X-30:* The Cold War SSTO

1. Newt Gingrich, *Window of Opportunity: A Blueprint for the Future* (New York: Tom Doherty Associates, Inc., 1984), p. 12.

2. Quoted in Scott Pace, "National Aero-space Plane Program: Principal Assumptions, Findings, and Policy Options," RAND publication P-7288-RGS, December 1986, p. 1.

3. Paul Czysz, interview by Erik M. Conway, tape and transcript, July 17, 2001, NASA Langley Research Center, Hampton, VA, pp. 1-5, 8-9, 11.

4. Russell J. Hannigan, *Spaceflight in the Era of Aero-Space Planes* (Malabar, FL: Krieger Publishing Company, 1994), p. 71. Materials in file 824, NASA Historical Reference Collection (hereafter NHRC), indicate that the article was Robert Goddard, "A New Turbine Rocket Plane for the Upper Atmosphere," *Popular Science,* December, 1931, pp. 148-149.

5. Irene Sänger-Bredt, "The Silver Bird Story: A Memoir," file 7,910, NHRC; Hannigan, pp. 71-73; Michael J. Neufeld, *The Rocket and the Reich: Peenemünde and the Coming of the Ballistic Missile Era* (New York: The Free Press, 1995), pp. 7-10; Richard P. Hallion, "In the Beginning Was the Dream. . .," in Hallion, ed., *The Hypersonic Revolution: Eight Case Studies in the History of Hypersonic Technology,* vol. 1, *From Max Valier to Project Prime, 1924-1967* (Dayton: Special Staff Office, Aeronautical Systems Division, Wright-Patterson AFB, 1987), pp. xi-xv.

6. Richard P. Hallion and James O. Young, "Space Shuttle: Fulfillment of a Dream," in Hallion, ed., *The Hypersonic Revolution: Eight Case Studies in the History of Hypersonic Technology,* vol. 2, *From Scramjet to the National Aero-Space Plane* (Dayton: Special Staff Office, Aeronautical Systems Division, Wright-Patterson AFB, 1987), p. 949

7. Hannigan, pp. 77-78.

8. Quoted in T. A. Heppenheimer, *The Space Shuttle Decision: NASA's Search for a Reusable Space Vehicle,* NASA SP-4221 (Washington, DC: Government Printing Office, 1999), p. 246.

9. On the closing of the ERC, see Boyd C. Myers II, *A Report on the Closing of the NASA Electronics Research Center, Cambridge, Massachusetts* (Washington, DC: NASA, October 1, 1970), especially Robert H. Rollins II, "Closing of the NASA Electronics Research Center: A Study of the Reallocation of Space Program Talent," pp. 106-187. Rollins wrote the study as his M.S. thesis at the Alfred P. Sloan School of Management, MIT.

10. Heppenheimer, *Shuttle,* pp. 252-253.

11. U.S. House of Representatives, Subcommittee on Space Science and Applications, Committee on Science and Technology, 98th Cong., 2d sess., *Review of Space*

Shuttle Requirements, Operations, and Future Plans, October 1984, serial GG (Washington, DC: Government Printing Office, 1984), pp. 3, 29, 30.

12. James E. Tomayko, *Computers Take Flight: A History of NASA's Pioneering Digital Fly-by-wire Project,* NASA SP-4224 (Washington, DC: NASA, 2000), pp. vii, ix-x, 17, 85, 87, 89, 108. For an understanding of the complexities of software programming back then, see ibid., pp. 90, 95-97.

13. Alan Wilhite, interview by author, tape and transcript, May 22, 1997, Langley Research Center, Hampton, VA, pp. 2, 5-7, NHRC; Charles H. "Chuck" Eldred, interview by author, tape and transcript, May 20, 1997, Langley Research Center, Hampton, VA, pp. 6, 9, NHRC; Delma "Del" Freeman, interview by author, tape and transcript, May 23, 1997, Langley Research Center, Hampton, VA, p. 3, NHRC; Raymond Chase, interview by author, tape and transcript, July 6, 2000, ANSER Corporation, Arlington, VA, pp. 1, 18-19, NHRC.

14. Rudolph C. Haefeli, Earnest G. Littler, John B. Hurley, and Martin G. Winter, Martin Marietta Corporation, Denver Division, *Technology Requirements for Advanced Earth-Orbital Transportation Systems: Final Report,* NASA Contractor Report CR-2866 (Washington, DC: NASA, October 1977), pp. 45, 75; Andrew K. Hepler and E. L. Bangsund, Boeing Aerospace Company, Seattle, WA, *Technology Requirements for Advanced Earth Orbital Transportation Systems,* vol. 1, *Executive Summary,* NASA Contractor Report CR-2878 (Washington, DC: NASA, 1978), p. 14.

15. Eldred, p 16.

16. Haefeli et al., pp. 117, 126, 178.

17. Hepler and Bangsund, 1:13-14, 21-22.

18. Andrew K. Hepler, interview by author, tape and transcript, July 11, 2000, Seattle, WA, pp. 13-15, NHRC; Eugene M. Emme, *Aeronautics and Astronautics: Chronology on Science, Technology, and Policy, 1915-1960* (Washington, DC: NASA, 1961), p. 17; Bill Gunston, *Rockets and Missiles* (London: Salamander, 1979), p. 39.

19. Roger E. Bilstein, *The American Aerospace Industry: From Workshop to Global Enterprise* (New York: Twayne Publishers, 1996), p. 83.

20. Hepler, pp. 2, 16; Hepler and Bangsund, 1:15.

21. Bilstein, pp. 178-179.

22. Hepler, p 4.

23. Hepler, pp. 5-6; Hepler and Bangsund, 1:15.

24. Boeing Aerospace Company, *Final Report on Feasibility Study of Reusable Aerodynamic Space Vehicle,* vol. 1, *Executive Summary* (Kent, WA: Boeing Aerospace Company, November 1976), pp. 5, 35.

25. Ibid., 1:2, 22.

26. Ibid., 1:31. I base this statement on the estimated cost of building a prototype for the air force in 1982, $1.4 billion. Boeing RASV proposal, December 1982, file 256, *X-33* Archive (see my bibliographic essay).

27. Boeing, *Final Report,* 1:25-27.

28. Ibid., 1:30; Hepler, p. 8; Chase, interview, pp. 3-4, 7-8, 11, 15.

29. Chase, interview, pp. 2, 10-11; Hepler, pp. 6, 9-11; Boeing Aerospace Company, *Final Report on Feasibility Study of Reusable Aerodynamic Space Vehicle,* vol. 2, *Vehicle Design and Analysis* (Kent, WA: Boeing Aerospace Company, November 1976), p. 218.

30. Boeing, *Final Report,* 2:216.

31. Hepler, p. 9.

32. Ibid., pp. 7, 10; Chase, interview, pp. 8-9.

33. Richard P. Hallion, "Yesterday, Today, and Tomorrow: From Shuttle to the National Aero-Space Plane," in Hallion, *Hypersonic Revolution,* 2:1334; P. Kenneth Pierpont, "Preliminary Study of Adaptation of SST Technology to a Reusable Aero-space Launch Vehicle System," NASA Langley Working Paper NASA-LWP-157, November 3, 1965; Boeing RASV proposal, December 1982, file 256, *X-33* Archive; Jess Sponable, interview by author, tape and transcript, January 16, 1998, NASA Headquarters, Washington, DC, pp. 4-5, 9, 18, NHRC; Gary Payton and Jess Sponable, "Designing the SSTO Rocket," *Aero-space America* 29, no. 4 (April 1991): 40.

34. R. L. Chase, "Science Dawn Overview," March 1990, file 235, *X-33* Archive; Sponable, interview, pp. 3-8; Sponable, comments on "X-Rocket" monograph, n.p., file 726, *X-33* Archive.

35. Sponable, interview, pp. 8-10; Major Stephen Clift, "Have Region Program: Final Brief," September 1989, file 235, *X-33* Archive; Sponable, comments on "X-Rocket" monograph, p. 41.

36. Hallion, "Yesterday, Today, and Tomorrow," 2:1336-1337.

37. Ibid., 2:1337, 1341.

38. Ibid., 2:1337, 1340.

39. Ibid., 2:1345.

40. John V. Becker, "Confronting Scramjet: The NASA Hypersonic Ramjet Experiment," in Hallion, 2:VI.xii, VI.xiv, 765, 786-789, 824, 841.

41. Heppenheimer, *The National Aerospace Plane* (Arlington, VA: Pasha Market Intelligence, 1987), p. 14; Hallion, "Yesterday, Today, and Tomorrow," 2:1361; Larry Schweikart, "The Quest for the Orbital Jet: The National Aerospace Plan Program, 1983-1995," pp. I.30-31, NHRC.

42. Hallion, "Yesterday, Today, and Tomorrow," 2:1346, 1351.

43. Becker, p. VI.xv; Hallion, "Yesterday, Today, and Tomorrow," 2:1362-1364.

44. Schweikart, pp. I.11, I.12.

45. Robert Jones, interview by Erik M. Conway, tape and transcript, June 25, 2001, NASA Langley Research Center, Hampton, VA, pp. 8-9; Conway to Butrica, e-mail, April 5, 2002.

46. Schweikart, pp. I.19-20, I.23, I.28, III.43.

47. Hallion, "Yesterday, Today, and Tomorrow," 2:1379; Schweikart, pp. III.31, III.43-44.

48. Ivan Bekey, interview by author, tape and transcript, March 2, 1999, NASA Headquarters, Washington, DC, p. 18, NHRC.

49. Schweikart, pp. I.22, I.29, I.30, II.9, VIII.1-3.

50. Ibid., p. I.25.

51. Ibid., pp. I.25-26, II.9, III.53, IV.31.

52. Hallion, "Yesterday, Today, and Tomorrow," 2:1379; Schweikart, pp. III.31, III.43-44.

53. Schweikart, pp. III.50, III.53, IV.3.

54. Ibid., pp. IV.21, IV.22, IV.24-25.

55. Ibid., pp. VI.33-36.

FIVE: Space Visionaries

1. Jules Verne, *De la terre à la lune. Trajet direct en 97 heures* (Paris: J. Hetzel, 1865); Verne, *The Baltimore Gun Club,* trans. Edward Roth (Philadelphia: King & Baird, 1874). The story originally appeared in serialized form in the *Journal des Débats politiques et littéraires* from September 14 through October 14, 1865.

2. Chrysler Corporation Space Division, Final Report on Project Single-stage Earth-orbital Reusable Vehicle Space Shuttle Feasibility Study, vol. 1. "Summary," TR-AP-71-4, June 30, 1971; Dennis R. Jenkins, *Space Shuttle: The History of Developing the National Space Transportation System* (Marceline, MO: Walsworth Publishing, 1992), p. 86.

3. T. A. Heppenheimer, *The Space Shuttle Decision: NASA's Search for a Reusable Space Vehicle,* NASA SP-4221 (Washington, DC: Government Printing Office, 1999), pp. 235-239.

4. Edward E. Gomersall, "A Single-Stage-to-Orbit Shuttle Concept," July 20, 1970, pp. 1-2, 8, 26, file 8,146, NASA Historical Reference Collection (hereafter NHRC).

5. Robert Salkeld and Rudi Beichel, "Reusable One-Stage-to-Orbit Shuttles: Brightening Prospects," *Astronautics & Aeronautics* 11, no. 6 (June 1973): 48.

6. Philip Bono and Kenneth Gatland, *Frontiers of Space,* rev. ed. (New York: Macmillan Publishing, 1976), pp. 140-142; G. Harry Stine, *Halfway to Anywhere: Achieving America's Destiny in Space* (New York: M. Evans and Co., 1996), pp. 40-41.

7. Bono and Gatland, pp. 173-183; Phil Bono, Frank E. Senator, and D. (Sam) Garcia, "The Enigma of Booster Recovery—Ballistic or Winged?" in Society of Automotive Engineers, *Space Technology Conference, May 9-12, 1967* (New York: Society of Automotive Engineers, 1967), pp. 57-71.

8. Bono and Gatland, pp. 197-200.

9. Stine, p. 44.

10. Michael A. G. Michaud, *Reaching for the High Frontier: The American Pro-Space Movement, 1972-1984* (New York: Praeger, 1986), pp. 255-256.

11. "Pacific American Space Ventures, Limited, A California Limited Partnership: Executive Summary," n.d., file 254, *X-33* Archive (see my bibliographic essay).

12. Gary C. Hudson to Thomas L. Kessler, "Comments on SSTO Briefing and a Short History of the Project," memorandum, December 17, 1990, p. 2, file 242, *X-33* Archive (hereafter, Hudson, "History"); Pacific American Launch Systems, "Phoenix Venture Plan," 1985, p. 10.1, file 254, *X-33* Archive.

13. Gary C. Hudson, "History of the Phoenix VTOL SSTO and Recent Developments in Single-Stage Launch Systems," December 1992, file 242, *X-33* Archive; Hudson, "Phoenix: A Commercial Reusable Single-Stage-to-Orbit Launch Vehicle," AAS 85-644, paper read at the Joint AAS/Japanese Rocket Society Symposium, December 15-19, 1985, Honolulu, HI, file 254, *X-33* Archive; Steve Hoeser, "Phoenix Launch System: System Summary," August 1986, file 254, *X-33* Archive; Pacific American Launch Systems, "Draft Technical Proposal," n.d., file 254, *X-33* Archive.

14. Pacific American Launch Systems, "Phoenix Venture Plan."

15. Steve Hoeser, comments on "X-Rocket" monograph, September 30, 1999, n.p., file 758, *X-33* Archive.

16. "An Evaluation of the Phoenix STS: Application to USAF Roles and Missions," 9

pp., n.d., file 254, *X-33* Archive; Pacific American Launch Systems, "Phoenix Briefing, USAF Space Division," February 1985, file 254, *X-33* Archive.

17. "Pacific American, Society Expeditions Sign Space Tourism Pact," *The Commercial Space Report* 9, no. 8 (September 1985): 1-2.

18. Hudson, "A Space Development Corporation," April 29, 1989, file 254, *X-33* Archive.

19. Hudson, "History," p. 2.

20. Max Hunter, interview by author, tape and transcript, June 19, 1998, San Carlos, CA, pp. 5-9, NHRC.

21. Max Hunter, "The Origins of the Shuttle (According to Hunter)," September 1972, p. 2, file 338, *X-33* Archive.

22. Hunter, interview, pp. 12-14, 18; Hunter materials in file 204, *X-33* Archive.

23. Hunter manuscripts in file 204, *X-33* Archive; Douglas Aircraft Company, Missiles and Space Systems Engineering, "RITA: The Reusable Interplanetary Transport Approach: An Informal Proposal," Douglas Report SM-38456, February 1961, file 253, *X-33* Archive.

24. Hunter, interview, pp. 10, 13-15, 18. For more on RITA, see Jenkins, pp. 22-23; Max Hunter, "Eve's Name Was Rita," August 13, 1990, file 204, *X-33* Archive.

25. Hunter, "Origins of the Shuttle," pp. 2-3.

26. Ibid., p. 3.

27. James E. Love and William R. Young, "Operational Experience of the X-15 Airplane as a Reusable Vehicle System," in Society of Automotive Engineers, *Space Technology Conference*, pp. 198-204.

28. Frank J. Dore, "Aircraft Design and Development Experience Related to Reusable Launch Vehicles," in George K. Chacko, ed., *Reducing the Cost of Space Transportation,* Proceedings of American Astronautical Society Seventh Goddard Memorial Symposium, March 4-5, 1969, Washington, DC (Washington, DC: American Astronautical Society, 1969), pp. 49-50.

29. Hunter, "Origins of the Shuttle," p. 3.

30. Star Clipper materials in file 338, *X-33* Archive; Hunter, interview, pp. 24-30; Heppenheimer, *Shuttle,* pp. 85-86; Joan Lisa Bromberg, *NASA and the Space Industry* (Baltimore: Johns Hopkins University Press, 1999), pp. 78-83, 86.

31. The paper is the 41-page manuscript Hunter, "ZENI," March 1964, presented to the Executive Secretary, National Aeronautics and Space Council, file 338, *X-33* Archive.

32. Hunter, "Origins of the Shuttle," p. 6.

33. Max Hunter to E. P. Wheaton, vice president for research and development, Lockheed, "Orbital Transportation," memorandum, October 28, 1965, pp. 1-2, file 338, *X-33* Archive.

34. Ibid., pp. 2-3.

35. Hunter, "Origins of the Shuttle," pp. 5-6.

36. Hunter to Wheaton, pp. 2-3.

37. Star Clipper materials in file 338, *X-33* Archive; Hunter, interview, pp. 24, 33-34; Heppenheimer, *Shuttle,* pp. 85-86, 245-246.

38. Max Hunter, "An Engineer's View of the Large Space Telescope," April 7, 1975; idem, "Strategic Dynamics and Space-Laser Weaponry," October 31, 1977; idem, "Musings on National Strategy," 1977-1979; and idem, "The Space Laser Battle Station:

Historical Perspective and Grand Strategy," November 1983, all in file 338, *X-33* Archive.

39. Hunter, interview, p. 35.

40. Max Hunter to Pat Ladner, "Some History of the SSTO Program as of Sept 13 1990," memorandum, September 13, 1990, n.p., file 242, *X-33* Archive (hereafter, Hunter, "History").

41. Hudson, "History," p. 2.

42. Hunter, "History," n.p.

43. The discussion of the *X-Rocket* and XOP that follows is based on *X-Rocket* and XOP briefing charts in file 241, *X-33* Archive.

44. Max Hunter, "The Opportunity," August 6, 1985, revised May 6, 1986, pp. 1, 2, file 204, *X-33* Archive.

45. Ibid., p. 3.

46. Ibid.

47. Gary Hudson, "X-Rocket and SSX," in Ray G. Miller, ed., *Making Orbit '94 White Paper* (Redondo Beach[?]: California Space Development Council, 1994), p. 6; Hudson, "History," p. 2.

48. Hunter, interview, pp. 35, 41; idem, "History," n.p.; Steve Hoeser to Pat Ladner, "Past SDIO Director's Perspective and Decision Base Related to the SSTO," memorandum for the record, December 7, 1990, p. 1, file 242, *X-33* Archive.

49. Hunter, "History," n.p.

50. Hunter, interview, p. 41; idem, "The SSX Single Stage Experimental Rocket," March 15, 1988, file 204, *X-33* Archive. See also idem, "The SSX—A True Spaceship," *The Journal of Practical Applications in Space* 1, no. 1 (Fall 1989): 41-62. This was the maiden issue of a journal backed by Daniel Graham.

51. Hunter, "SSX," 1988, pp. 9, 11.

52. Ibid., pp. 16, 18-19.

SIX: Launching the *SSX*

1. Trudy E. Bell, "American Space-Interest Groups," *Star & Sky* 2 (September 1980): 54-55; Michael A. G. Michaud, *Reaching for the High Frontier: The American Pro-Space Movement, 1972-1984* (New York: Praeger, 1986), pp. 76-77, 137.

2. Today, the organization calls itself the L5 Society. I have kept the original L-5 spelling throughout, with the exception of the newsletter, which was always the *L5 News.*

3. Michaud, pp. 95, 125-126.

4. Michaud, pp. 89-93; Bell, "American Space-Interest Groups," p. 54.

5. Trudy E. Bell, "Upward: Status Report and Directory of the American Space Interest Movement, 1984-1985," photocopy, printed February 1985, pp. 10, 11, 17, file 290, *X-33* Archive (see my bibliographic essay).

6. Michaud, pp. 171-174.

7. Michaud, pp. 227-228. See also Daniel Graham, *The Non-Nuclear Defense of Cities: The High Frontier Space-Based Defense Against ICBM Attack* (Cambridge, MA: Abt Books, 1983).

8. Daniel Graham, *High Frontier: A New National Strategy* (Washington, DC: The Heritage Foundation, 1982).

9. Bell, "Upward," p. 42.

10. Jerry Pournelle, "Draft Response," December 16, 1998, pp. 1-2, file 754, *X-33* Archive. A rather different view of the Citizens' Advisory Council's origins is given by cofounder and science fiction writer G. Harry Stine. See his *Halfway to Anywhere: Achieving America's Destiny in Space* (New York: M. Evans and Co., 1996), pp. 76-77. Pournelle, "Draft Response," p. 1, countered that the meeting Stine describes did take place, but was a meeting of L-5 Society officers.

11. Jerry Pournelle to Mark Albrecht, July 24, 1989, file 252, *X-33* Archive.

12. Pournelle, "Draft Response," pp. 2-3.

13. Ibid.

14. Michaud, pp. 128-129, 169; "Citizens' Advisory Council on National Space Policy, Jerry E. Pournelle, Ph.D., Chairman, August 10, 1997 Meeting," *http://www.jerrypournelle.com/slowchange/Citizen.html.*

15. Bell, "Upward," p. 101; Bell, "American Space-Interest Groups," p. 60; Michaud, pp. 124-127; William Sims Bainbridge, "The Science Fiction Subculture," in *The Spaceflight Revolution: A Sociological Study* (New York: John Wiley & Sons, 1976), pp. 198-234.

16. Indeed, according to Pournelle's website, the Citizens' Advisory Council "reported directly to the National Security Advisor through 1988." *http://www.jerrypournelle.com/slowchange/Citizen.html.*

17. Pournelle, "Draft Response," p. 3.

18. According to Michaud, pp. 170-171.

19. Goldin was not a member of the Citizens' Advisory Council. In honor of the *Mars Pathfinder* mission, Marilyn Niven baked Goldin a "Mars Cake." Photographs of Goldin with the "Mars Cake" and Goldin with Larry Niven and Jerry Pournelle and their wives at the August 10, 1997, meeting can been seen at: *http://www.jerrypournelle.com/slowchange/Citizen.html.*

20. Stine, p. 79.

21. Citizens' Advisory Council on National Space Policy, *America: A Spacefaring Nation Again* (Tucson: The L-5 Society, 20 July 1986), pp. 11, 33 (hereafter, *A Spacefaring Nation).*

22. Ibid., pp. 11-12, 24.

23. Ibid., p. 45.

24. James Muncy, interview by author, tape and transcript, January 12, 1999, Washington, DC, NASA Historical Reference Collection (hereafter NHRC), p. 4.

25. *A Spacefaring Nation,* pp. 49, 55.

26. Ibid., p. 51.

27. Internal evidence in *A Spacefaring Nation,* p. 7, which reproduces the letter sent to President Reagan, dated July 20, 1986.

28. *A Spacefaring Nation,* p. 24.

29. Stine, pp. 79, 80.

30. Steve Hoeser to Pat Ladner, "Addendum to Max's SSTO Program History," memorandum, September 24, 1990, p. 2, file 242, *X-33* Archive, gives the date of the meeting. Daniel O. Graham, *Confessions of a Cold Warrior: An Autobiography* (Fairfax, VA: Preview Press, 1995), p. 205, claims the idea to hold the meeting was his. Pournelle, as head of the Council, would have convened the meeting.

31. Gary C. Hudson to Thomas L. Kessler, "Comments on SSTO Briefing and a

Short History of the Project," memorandum, December 17, 1990, p. 3, file 242, *X-33* Archive (hereafter, Hudson, "History").

32. Larry Niven, Jerry Pournelle, and Michael Flynn, *Fallen Angels* (Riverdale, NY: Baen Books, 1991), republished by Mass Market Paperback, 1992. The reference to this novel is from Steve Hoeser, interview by author, tape and transcript, January 15, 1999, Boeing, Crystal City, VA, pp. 49-50, NHRC.

33. Pournelle, "Draft Response," p. 4.

34. Max Hunter to Pat Ladner, "Some History of the SSTO Program as of Sept 13 1990," memorandum, September 13, 1990, n.p., file 242, *X-33* Archive (hereafter Hunter, "History").

35. Pournelle, "Draft Response," p. 5.

36. Graham, *Confessions,* pp. 203, 204.

37. Hudson, "History," p. 3. Actually, prior to the meeting, *SSX* stood for *Single Stage eXperimental.* It was at that meeting, though, that the name change to *SpaceShip* or *Space Ship eXperimental* took place, according to Hunter, "History," n.p.

38. Stine, pp. 79, 81. The "white paper and supporting technical documentation" probably are "SSX and Current Space Operations," 1989, and "SSX (Spaceship Experimental)," 1989, in file 231, *X-33* Archive.

39. Graham, *Confessions,* p. 202.

40. Ibid., pp. 203, 204.

41. Max Hunter, interview by author, tape and transcript, June 19, 1998, San Carlos, CA, p. 41, NHRC; idem, "The SSX Single Stage Experimental Rocket," March 15, 1988, file 204, *X-33* Archive (hereafter Hunter, "SSX," 1988); idem, "The SSX—A True Spaceship," *The Journal of Practical Applications in Space* 1, no. 1 (Fall 1989): 41-62.

42. Hunter, "History," n.p.

43. Hunter, "SSX," 1988, pp. 9, 11, 16, 18-19.

44. Hunter, "History," n.p. The same story is consistent with other accounts, such as Hudson, "History," p. 3.

45. Graham, *Confessions,* p. 205.

46. "The Quayle Presentation Charts; Comments by MWH [Maxwell W. Hunter]," n.d., file 240, *X-33* Archive.

47. "On Getting Administration Support for the SSX," file 240, *X-33* Archive.

48. Ibid.

49. Hunter, "History," n.p.

50. Steve Hoeser to Pat Ladner, "Past SDIO Director's Perspective and Decision Base Related to the SSTO," memorandum for the record, December 7, 1990, file 242, *X-33* Archive.

51. Graham, *Confessions,* p. 205.

52. The agency was known at its inception in 1958 as the Advanced Research Projects. Beginning in 1972, its name was the Defense Advanced Research Projects Agency (DARPA), before it reverted to ARPA in 1993. Today, it is again DARPA. For the sake of clarity and simplicity, I have used ARPA throughout.

53. Graham, *Confessions,* p. 205.

54. "On Getting Administration Support for the SSX."

55. Graham, *Confessions,* p. 205.

56. Hunter, "History," n.p. Graham, *Confessions,* p. 205, gives the date as February 12, 1989, however.

57. Graham, *Confessions,* p. 205.

58. Graham, "Briefing for Vice President Quayle," pp. 1-3, file 240, *X-33* Archive; Graham, *Confessions,* p. 205.

59. Hunter, "History," n.p.

60. Graham, "Briefing for Vice President Quayle," p. 4. A portion of the material quoted here also appeared in Stine, p. 81

61. Graham, *Confessions,* p. 205.

62. Jerry Pournelle to J. Danforth Quayle, February 27, 1989, file 252, *X-33* Archive.

63. Idem, July 24, 1989, file 252, *X-33* Archive.

64. Graham, *Confessions,* p. 205.

65. J. Danforth Quayle, *Standing Firm: A Vice-Presidential Memoir* (New York: HarperCollins Publishers, 1994). The quotation is from p. 180.

66. Hunter, "History," n.p.

67. Graham to Robert M. Guttman, February 21, 1989, file 252, *X-33* Archive.

68. For example, see "The Experimental Spaceship (SSX)," *High Frontier Newsletter* 6, no. 12 (December 1989): 1.

69. Hunter, "History," n.p.; Hoeser, interview, January 15, 1999, p. 51.

70. Hoeser, interview, January 15, 1999, pp. 46-47.

71. Monahan succeeded Abrahamson on February 1, 1989. Edward Reiss, *The Strategic Defense Initiative* (New York: Cambridge University Press, 1992), pp. 180-181; Donald R. Baucom, "Ballistic Missile Defense: A Brief History," found at the BMDO website *www.acq.osd.mil/bmdo/bmdolink/html/origins.html.*

72. Graham, *Confessions,* pp. 205-206.

73. Baucom, "Ballistic Missile Defense."

74. Steve Hoeser, interview by author, tape and transcript, February 11, 1999, Boeing, Crystal City, VA, NHRC, p 7.

75. Steve Hoeser to Pat Ladner, "Addendum to SSTO Program History," memorandum, September 24, 1990, file 242, *X-33* Archive.

76. Jay P. Penn, C. L. Leonard, and C. A. Lindley, "Review of Pacific American Launch System SSX:Phoenix VTOL Concept," July 19, 1989, p. 2, file 255, *X-33* Archive.

77. Hudson, "History," p. 3.

78. Max Hunter to Jess Sponable, October 1, 1989, file 252, *X-33* Archive; Jay Penn, comments on "X-Rocket" monograph, June 28, 1999, p. 38, file 755, *X-33* Archive.

79. Jess Sponable, interview by author, tape and transcript, January 16, 1998, NASA Headquarters, Washington, DC, p. 13, NHRC.

80. Ibid., p. 14; Jess Sponable to Jim Ransom, July 26, 1989, file 252, *X-33* Archive.

81. Daniel Graham to George L. Monahan, Jr., August 7, 1989, file 252, *X-33* Archive.

82. Larry Schweikart, "The Quest for the Orbital Jet: The National Aerospace Plan Program, 1983-1995," pp. VI.3, VI.5, NHRC.

83. Sponable, interview, p. 3.

84. Jerry Pournelle to Stewart Nozette, July 24, 1989, file 252, *X-33* Archive.

85. Sponable, interview, p 14.

86. Gary E. Payton, interview by author, tape and transcript, August 20, 1997, NASA Headquarters, Washington, DC, p. 25, NHRC.

87. Payton, comments on "X-Rocket" monograph, June 28, 1999, p. 38, file 725, *X-33* Archive.

88. See, for example, Graham to Monahan.

89. Penn, Leonard, and Lindley, p. 1.
90. Pournelle to Quayle, July 24, 1989.

SEVEN: The SDIO SSTO Program

1. Steve Hoeser, "SSX Update and Options," 12-14 August 1989, file 252, *X-33* Archive (see my bibliographic essay).
2. George L. Monahan, Jr., to Daniel Graham, January 18, 1990, file 252, *X-33* Archive.
3. Vincent Kiernan, "Budget Crunch Bumps SSTO Tests," *Space News,* March 18-24, 1991, p. 3.
4. Michael Griffin, interview by author, tape and transcript, August 18, 1997, Orbital Sciences Corporation, Dulles, VA, p. 22, NASA Historical Reference Collection (hereafter NHRC); Monahan to Graham: "I have assigned our Technology Deputate, Dr. Michael Griffin and Colonel Gary Payton, the task of managing this entire process."
5. Steve Hoeser to Pat Ladner, "Addendum to SSTO Program History," memorandum, September 24, 1990, file 242, *X-33* Archive; Steve Hoeser, interview by author, tape and transcript, February 11, 1999, Boeing Large System Integrator, Alexandria, VA, p. 2, NHRC. Daniel O. Graham, *Confessions of a Cold Warrior: An Autobiography,* (Fairfax, VA: Preview Press, 1995), p. 205, also refers to this briefing.
6. Hoeser to Ladner, "Addendum to SSTO Program History."
7. Griffin, interview, pp. 7, 10-11, 15.
8. Gary E. Payton, interview by author, tape and transcript, August 20, 1997, NASA Headquarters, Washington, DC, pp. 7-8, NHRC.
9. Monahan to Graham.
10. "Single Stage Experiment for the Technology Deputy," *Commerce Business Daily,* March 26, 1990, issue PSA-0057, p. 2, col. 1, photocopy, file 247, *X-33* Archive.
11. Griffin, interview, p. 22.
12. Gary E. Payton, comments on "X-Rocket" monograph, undated, p. 46, file 725, *X-33* Archive.
13. Payton, interview, pp. 28-29.
14. Griffin, interview, p. 23.
15. Phase I Statement of Work, p. 2, file 257, *X-33* Archive. A description of the SSTO Program also is in Gary Payton and Jess Sponable, "Designing the SSTO Rocket," *Aero-space America* 29, no. 4 (April 1991): 36-39.
16. Griffin, interview, p. 23.
17. "SSTO Pre-Proposal Conference Questions/Answers," p. 6, in "Pre-proposal Conference Single Stage to Orbit Concept Evaluation for the SDIO Technology Directorate," May 16, 1990, file 257, *X-33* Archive.
18. Jess Sponable, comments on "X-Rocket" monograph, p. 48, file 726, *X-33* Archive.
19. Phase I Statement of Work, p. 2.
20. Gary C. Hudson to Thomas L. Kessler, "Comments on SSTO Briefing and a Short History of the Project," memorandum, December 17, 1990, p. 3, file 242, *X-33* Archive (hereafter, Hudson, "History").
21. Jess Sponable, comments on Andrew J. Butrica, "The Spaceship That Came in

from the Cold War" (unpublished manuscript), December 2000, p. 357, file 856, *X-33* Archive.

22. Phase I Statement of Work, pp. 3, 4.

23. Gary Payton, comments on "X-Rocket" monograph, February 1, 1999, n.p., file 725, *X-33* Archive., n.p.

24. Ibid.

25. Ibid., p. 50. Payton later would date the announcement of the X-33 Phase II award also to avoid the July 4 holiday.

26. "Pre-proposal Conference." The entire following section is from this document and the Phase I Statement of Work.

27. This was the same general configuration that Martin Marietta and Boeing had studied and dismissed in NASA Langley studies undertaken in the 1970s.

28. McDonnell Douglas Space Systems Company, "Single Stage to Orbit Program Phase I Concept Definition," December 13, 1990, file 267, *X-33* Archive; General Dynamics Space Systems Division, "Concept Review Technical Briefing," December 13, 1990, file 265, *X-33* Archive; Boeing Defense and Space Group, Space Transportation Systems, "Single Stage to Orbit Technology Demonstration Concept Review Technical Briefing," December 12, 1990, file 264, *X-33* Archive; Rockwell International, "SDIO Single Stage to Orbit Concept Review," December 12, 1990, file 259, *X-33* Archive.

29. Jay Penn to Lt. Col. Pat Ladner, January 21, 1991, file 255, *X-33* Archive.

30. Jay Penn, "Comments on the SDIO Program," n.p., file 255, *X-33* Archive.

31. Ibid.

32. Ibid.

33. Michael D. Griffin to Richard H. Truly, February 22, 1991, file 294, *X-33* Archive.

34. Arnold D. Aldrich to Michael D. Griffin, March 14, 1991, file 294, *X-33* Archive.

35. "NASA Evaluation of SDIO Phase I SSTO Concepts," n.d., file 294, *X-33* Archive.

36. Boeing Aerospace Co., "SSTO Technology Demonstration Final Report and Preferred Concept Design Report," June 1991, file 263, *X-33* Archive.

37. NASA Langley Research Center, Vehicle Analysis Branch, "NASA Review of the SDIO Phase I Single-Stage-to-Orbit Concepts," n.d., but took place in the first half of 1991, file 294, *X-33* Archive.

38. Doug Stanley, interview by author, tape and transcript, February 25, 1999, Orbital Sciences Corporation, Dulles, VA, p. 13, NHRC.

39. General Dynamics Space Systems Division, "Single-Stage-to-Orbit (SSTO) Technology Demonstration Program Final Review," May 3, 1991, file 266, *X-33* Archive.

40. "SSTO Design Update Presented to Langley Research Center, April 16, 1991," file 260, *X-33* Archive; Rockwell International, "SDIO Single Stage to Orbit Phase I Program Review," May 2, 1991, file 262, *X-33* Archive.

41. Cover Letter, Phase II RFP, file 269, *X-33* Archive.

42. Phase II Statement of Work, p. 4, file 269, *X-33* Archive.

43. Ibid., pp. 5, 7, 9.

44. Ibid., pp. 2, 5, 6, 8, 9.

45. Ibid., pp. 5-6.

46. Ibid., pp. 6, 10-11.

47. Ibid., pp. 3, 4.

48. General Dynamics, "Single-Stage-to-Orbit (SSTO) Technology Demonstration Program Final Review"; Rockwell International, "SDIO Single Stage to Orbit Phase I Program Review"; McDonnell Douglas Space Systems Company, "Single Stage to Orbit Preferred Concept Design Report," June 15, 1991, p. 1, file 268, *X-33* Archive.

49. Griffin, interview, p. 25.

50. Jess Sponable, interview by author, tape and transcript, January 16, 1998, NASA Headquarters, Washington, DC, pp. 18, 19, NHRC.

51. Ibid., p. 17.

52. Maj. Michael F. Doble, U.S. Army, Chief, News Branch, fax, announcement of award, August 19, 1991, file 291, *X-33* Archive.

53. William Gaubatz, interview by author, tape and transcript, October 25, 1997, Huntington Beach, CA, p. 33, NHRC.

54. For a discussion of rapid prototyping, see John Bruce, *Rapid Prototyping and Manufacturing* (Norwalk, CT: Business Communications, 1993) and Chua Chee Kai and Leong Kah Fai, *Rapid Prototyping: Principles and Applications in Manufacturing* (New York: Wiley, 1996).

55. *Rapid Prototyping Systems: Fast Track to Product Realization: A Compilation of Papers from Rapid Prototyping and Manufacturing '93* (Dearborn, MI: Society of Manufacturing Engineers in Cooperation with Rapid Prototyping Association of SME, 1994); Rolf-Jürgen Ahlers and Gunther Reinhart, eds., *Rapid Prototyping: 10-11 June, 1996, Besançon, France* (Bellingham, WA: SPIE, 1996); idem, *Rapid Prototyping and Flexible Manufacturing: 16 June 1997, Munich, Germany* (Bellingham, WA: SPIE, 1997); IEEE International Workshop on Rapid System Prototyping, *Proceedings: Shortening the Path from Specification to Prototype* (Los Alamitos, CA: IEEE Computer Society, 1995).

56. Paul L. Klevatt, interview by author, tape and transcript, July 14, 2000, Tustin, CA, pp. 8-10, NHRC; William Gaubatz, "Rapid Prototyping," in *Proceedings of the IEEE Aerospace Applications Conference, February 3-10, 1996* (Piscataway, NJ: IEEE Press, 1997), vol. 3, pp. 303-6; idem, comments on "X-Rocket" monograph, pp. 62, 64, file 751, *X-33* Archive.

57. Charles "Pete" Conrad, interview by author, tape and transcript, October 22, 1997, Rocket Development Company, Los Alamitos, CA, p. 5, NHRC.

58. See, for example, the narrative by Ben R. Rich and Leo Janos, *Skunk Works: A Personal Memoir of My Years at Lockheed* (Boston: Little, Brown, 1994), as well as the less-biased work by Steve Pace, *Lockheed Skunk Works* (Osceola, WI: Motorbooks International, 1992), and Jay Miller, *Lockheed Martin's Skunk Works* (North Branch, MN: Specialty Press Publishers & Wholesalers, 1995).

59. Rich and Janos, pp. 51-53, 111-112.

60. Max Hunter, "The Weasel Works," July 4, 1992, file 204, *X-33* Archive.

61. Gaubatz, comments on "X-Rocket" monograph, p. 62.

62. Max Hunter, interview by author, tape and transcript, June 19, 1998, San Carlos, CA, pp. 18, 30, NHRC; idem, comments on "X-Rocket" monograph, p. 62, file 757, *X-33* Archive.

63. Gaubatz, "Rapid Prototyping," 3:303-306; idem, comments on "X-Rocket" monograph, pp. 62, 64.

64. Much has been written about Deming, especially by his "apostles." See, for example, Nancy R. Mann, *The Keys to Excellence: The Story of the Deming Philosophy* (Los

Angeles: Prestwick Books, 1985); Mary Walton, *The Deming Management Method,* foreword by W. Edwards Deming (New York: Dodd, Mead, 1986); Andrea Gabor, *The Man Who Discovered Quality: How W. Edwards Deming Brought the Quality Revolution to America: The Stories of Ford, Xerox, and GM* (New York: Times Books, 1990); Rafael Aguayo, *Dr. Deming: The American Who Taught the Japanese About Quality* (Secaucus, NY: Carol Publishing Group, 1990); and Frank Voehl, ed., *Deming: The Way We Knew* (Delray Beach, FL: St. Lucie Press, 1995).

65. Deming, *Out of the Crisis* (Cambridge: MIT Press, 1986), pp. 23-90.

66. A useful source is John Sheldrake, *Management Theory: From Taylorism to Japanization* (Boston: International Thomson Business Press, 1997).

67. This is the thesis put forth by Alfred DuPont Chandler, *The Visible Hand: The Managerial Revolution in American Business* (Cambridge, MA: Belknap Press, 1977).

68. On Edison's West Orange laboratory and manufacturing plant, see Andre Millard, Duncan Hay, and Mary K. Grassick, *Edison Laboratory: Edison National Historic Site, West Orange, New Jersey* (West Orange, NJ: Division of Historic Furnishings, Harper's Ferry Center, National Park Service, 1995).

69. Gaubatz, interview, p. 6.

70. Klevatt, interview, p.14; William Gaubatz and Paul Klevatt, "MDSSC Guidelines for Single-Stage-to-Orbit Rapid Prototyping Department," September 1991, pp. 1-5, file 261, *X-33* Archive (hereafter Gaubatz and Klevatt).

71. Gaubatz, interview, p. 62; Klevatt, interview, pp. 7-9; Mark A. Gottschalk, "Delta Clipper: Taxi to the Heavens," *Design News,* September 1992, n.p., photocopy, file 292, *X-33* Archive; Leonard David, "Unorthodox New DC-X Rocket Ready for First Tests," *Space News,* January 11-17, 1993, p. 10.

72. Luis Zea, "The Quicker Clipper," *Final Frontier,* October 1992, p. 4; McDonnell Douglas Space Systems Company, "Single Stage to Orbit Program Phase I Concept Definition."

73. Penn, "Comments on the SDIO Program."

74. Sponable, comments on "X-Rocket" monograph, p. 67.

75. Gaubatz, comments on "X-Rocket" monograph, p. 67.

76. Gaubatz, interview, p. 32.

77. Payton and Sponable, p. 43. See also Virginia P. Dawson with Mark D. Bowles, *The Development of Centaur: Upper Stage and American Rocketry* (Washington, DC: NASA, 2003).

78. Klevatt, interview, pp. 14-15. According to Jess Sponable, the air force did not own the engines. They came from the Pratt & Whitney production line before the government bought them, so they were actually the property of Pratt & Whitney. Sponable, comments on "X-Rocket" monograph, p. 67. Klevatt, interview, p. 15, confirms this.

79. Klevatt, interview, pp. 16, 18, 20; Zea, p. 4; McDonnell Douglas Space Systems Company, "Single Stage to Orbit Program Phase I Concept Definition."

80. The rectangular-shaped, 2219 aluminum hydrogen tank was 8 feet (2.4 m) in diameter and 16 feet (about 4.9 m) long. The conical-shaped, 2219 aluminum oxygen tank was 7 feet 10 inches (about 2.4 m) in diameter at its widest point and 9 feet (2.7 m) long. Paul Klevatt, comments on Butrica, "The Spaceship That Came in from the Cold War," October 18, 2000, p. 2, file 858, *X-33* Archive.

81. Klevatt, interview, pp. 17, 21; Zea, p. 4; McDonnell Douglas Space Systems Company, "Single Stage to Orbit Program Phase I Concept Definition."

82. Payton, comments on "X-Rocket" monograph, p. 68.

83. Gaubatz, comments on "X-Rocket" monograph, p. 68.

84. Gottschalk.

85. Conrad, interview, p. 3.

86. Ibid., pp. 4, 9.

87. John C. Lonnquest and David F. Winkler, *To Defend and Deter: The Legacy of the United States Cold War Missile Program,* USACERL Special Report 97/01 (Champaign, IL: U.S. Army Construction Engineering Research Laboratories, 1996), pp. 132-133.

88. Donald R. Baucom, "Ballistic Missile Defense: A Brief History," found at the BMDO website: *www.acq.osd.mil/bmdo/bmdolink/html/origins.html.*

89. Donald R. Baucom, *The Origins of SDI, 1944-1983* (Lawrence: University Press of Kansas, 1992), p. 199; idem, "Ballistic Missile Defense: A Brief History."

90. Vincent Kiernan, "Firms on SDI Rollercoaster," *Space News,* August 24-30, 1992, p. 21. The price per copy was to be $1 million.

91. Kiernan, "Budget Crunch Bumps SSTO Tests," p. 28.

92. Lyn Ragsdale, "Politics Not Science: The U.S. Space Program in the Reagan and Bush Years," in Roger D. Launius and Howard E. McCurdy, eds., *Spaceflight and the Myth of Presidential Leadership* (Urbana: University of Illinois Press, 1997), p. 161.

93. Ibid., pp. 163-164.

94. William C. Berman, *America's Right Turn,* 2d ed. (Baltimore: Johns Hopkins University Press, 1998), p. 121.

95. Ragsdale, p. 154.

96. J. Danforth Quayle, *Standing Firm: A Vice-Presidential Memoir* (New York: HarperCollins Publishers, 1994), pp. 179, 180.

97. Ibid., pp. 179, 180.

98. Ibid., pp. 181, 184, 189-190.

99. Ibid., pp. 181, 185, 190.

100. "Pre-Hearing Questions Submitted to Daniel S. Goldin Submitted by the Majority," n.d., file 4511, NHRC; Craig Covault, "Nominee for NASA Chief Fits Space Council Approach," *Aviation Week & Space Technology* 136, no.11 (March 16, 1992): 21; "Daniel S. Goldin" Data for one-page news release; and David C. Morrison, "NASA's Big Bang," *Government Executive,* February 1993, p. 18, file 4512, NHRC.

101. J. Danforth Quayle, "Prepared Remarks of the Vice President to the American Institute of Aeronautics and Astronautics," May 1, 1990, p. 10, file 513, *X-33* Archive.

EIGHT: W(h)ither SSTO?

1. Maj. Michael F. Doble, U.S. Army, Chief, News Branch, fax, announcement of award, August 19, 1991, file 291, *X-33* Archive (see my bibliographic essay); "Solicitation for the Single Stage to Orbit (SSTO) Concept Evaluation for the Technology Directorate (SDIO/TN), RFP no. SDIO84-90-R-0007," cover letter, file 257, *X-33* Archive.

2. Budget figures are from Roger Handberg, *The Future of the Space Industry: Private Enterprise and Public Policy* (Westport, CT: Quorum Books, 1995), p. 86.

3. David N. Spires, *Beyond Horizons: A Half Century of Air Force Space Leadership*

(Peterson AFB: Air Force Space Command, 1997), pp. 227, 237, 268, 276; David P. Radzanowski to Rep. Dana Rohrabacher, "U.S. Launch Vehicle Development Programs," memorandum, May 7, 1993, pp. 3-6, file 444, *X-33* Archive.

4. Larry Schweikart, "The Quest for the Orbital Jet: The National Aerospace Plane Program, 1983-1995," p. VII.4, NASA Historical Reference Collection (hereafter NHRC).

5. William Gaubatz, interview by author, tape and transcript, October 25, 1997, Huntington Beach, CA, p. 35, NHRC.

6. Steve Hoeser to Pat Ladner, "Addendum to SSTO Program History," memorandum, September 24, 1990, pp. 3-4, file 242, *X-33* Archive; Max Hunter to Pat Ladner, "Some History of the SSTO Program as of Sept 13 1990," memorandum, September 13, 1990, n.p., file 242, *X-33* Archive (hereafter Hunter, "History"). Documents in file 251, *X-33* Archive, suggest the range of House and Senate committees of the 101st Congress considered for briefings: the Committee on Appropriations, the Committee on Armed Services, the Committee on Commerce, Science, and Transportation, and the Committee on the Budget in the Senate; the Committee on Science, Space, and Technology, the Committee on the Budget, and the Committee on Appropriations in the House.

7. Gaubatz, interview, pp. 12, 17; Daniel O. Graham, *Confessions of a Cold Warrior: An Autobiography* (Fairfax, VA: Preview Press, 1995), p. 209.

8. Jess Sponable, interview by author, tape and transcript, January 16, 1998, NASA Headquarters, Washington, DC, p. 20, NHRC.

9. "SSX Logbook," file 258, *X-33* Archive.

10. Graham, *Confessions,* p. 207.

11. James Muncy, interview by author, tape and transcript, January 12, 1999, Washington, DC, pp. 19-20, NHRC; "Space Frontier Foundation," file 344, *X-33* Archive.

12. "Space Access Society," file 238, *X-33* Archive.

13. Various documents in file 447, *X-33* Archive.

14. Sponable, interview, p. 21.

15. Speeches and other documents in file 291, *X-33* Archive.

16. Gaubatz, interview, p. 38.

17. Tim Kyger, interview by author, tape and transcript, February 12, 1999, Universal Space Lines, Arlington, VA, pp. 65-68, NHRC; Muncy, interview, p. 22; Kyger, comments on "X-Rocket" monograph, n.p., file 753, *X-33* Archive.

18. Kyger, interview, p. 84.

19. Kyger, "SSTO Phone Script & Talking Points," September 23, 1991; idem, "SDIO's SSTO Program: What We Want," n.d., file 291, *X-33* Archive.

20. Richard Cheney to Rep. Dana Rohrabacher, December 2, 1991, file 291, *X-33* Archive.

21. Charles Ordahl to Tim Kyger, March 27, 1992, file 291, *X-33* Archive.

22. Dana Rohrabacher to Samuel Skinner, Chief of Staff to the President, February 7, 1992, file 291, *X-33* Archive.

23. Daniel O. Graham to J. Danforth Quayle, February 21, 1992, file 291, *X-33* Archive.

24. Donald C. Fraser to Dana Rohrabacher, April 8, 1992, file 291, *X-33* Archive.

25. Tim Kyger, "Memorandum," April 30, 1992, p. 3, file 291, *X-33* Archive.

26. Ibid., pp. 3-8.

27. Ibid., p. 8.

28. Schweikart, p. VII.40.

29. Kyger, "Memorandum," pp. 9-10.

30. Ibid., pp. 8-9.

31. Tim Kyger, "Defense Agencies: Single-stage-to-orbit (SSTO)," May 14, 1992, and "Draft Report Language for Inclusion in the FY '93 DoD Authorization Bill (S.2629)," July 16, 1992, Binder 1, file 759, *X-33* Archive.

32. J. Milnor Roberts to unknown recipient, High Frontier letterhead, June 30, 1992, file 291, *X-33* Archive; Danielle Herubin, "House Axes Delta Clipper Funds," *The Orange County Register,* June 30, 1992, p. C2, photocopy, file 291, *X-33* Archive.

33. Phase II Statement of Work, p. 9, file 269, *X-33* Archive; Henry F. Cooper to Dana Rohrabacher, June 23, 1992, file 291, *X-33* Archive; "SSRT PDR Sign In," June 30–July 2, 1992, file 291, *X-33* Archive.

34. Tim Kyger to American Law Division, Congressional Research Service, August 25, 1992, and idem to "To whom it may concern," July 22, 1992, Binder 1, file 759, *X-33* Archive.

35. Space Frontier Foundation, press release, July 2, 1992, file 291, *X-33* Archive.

36. Muncy, interview, p. 28.

37. Jess Sponable, comments on "X-Rocket" monograph, p. 89, file 726, *X-33* Archive.

38. Beth Schwinn, "Funding Restored for Delta Clipper," *The Orange County Register,* July 3, 1992, p. B5, photocopy, file 291, *X-33* Archive.

39. Vincent Kiernan, "Rep. Dana Rohrabacher: Newsmaker Forum," *Space News,* August 17-23, 1992, p. 22.

40. "SSTO Gets Launch, But No Orbit," *Space News,* October 12-18, 1992, p. 2.

41. Henry F. Cooper to Rep. John P. Murtha, July 1, 1992, file 441, *X-33* Archive; Kyger, "Memorandum," p. 2; Space Transportation Association, "Single Stage to Orbit Vehicles: Near Term or Long Term?" August 1992, p. 3, file 291, *X-33* Archive.

42. James R. Asker, "SDI Organization Plans 1994 Test Flight of Single-Stage-to-Orbit Spacecraft," *Aviation Week & Space Technology* 133, no. 19 (November 5, 1990): 27.

43. Tim Kyger, "SSTO Phase III: Where Should It Reside?" May 26, 1992, p. 3, file 291, *X-33* Archive.

44. Ibid.

45. Ibid., pp. 4, 6; Kyger, comments on "X-Rocket" monograph, p. 88.

46. Kyger, "SSTO Phase III," pp. 3, 4.

47. "Single-Stage-to-Orbit Project Looks for New Home," *Space Commerce Week,* August 14, 1992, p. 2.

48. Space Transportation Association, "Single Stage to Orbit Vehicles," August 1992, cover page.

49. Space Transportation Association, "Single Stage to Orbit Vehicles: Near Term or Long Term?" draft, July 28, 1992, p. 9, file 291, *X-33* Archive.

50. Space Transportation Association, "Single Stage to Orbit Vehicles," August 1992, p. 11.

51. Kyger, comments on "X-Rocket" monograph, p. 90.

52. Dana Rohrabacher et al. to Joseph M. McDade, Subcommittee on Defense, House Committee on Appropriations, September 7, 1993; Robert Dornan, Dana Rohrabacher, Norman Mineta, and Clarence Brown to John M. Deutch, October 21,

1993; and Lt. Gen. Malcolm R. O'Neill to Rohrabacher, November 24, 1993, both in file 293, *X-33* Archive.

53. John Cunningham, "Sabotaging SSRT's Future," *Space News,* October 18-24, 1993, p. 24.

54. Space Access Society, *Space Access Update #21,* September 27, 1993, file 238, *X-33* Archive.

55. Dana Rohrabacher to Gary L. Denman, Director, ARPA, December 22, 1993, file 293, *X-33* Archive.

NINE: The Disorder of Things

1. Ben Iannotta, "SDIO Renamed, Deutch Gets Oversight," *Space News,* May 17-23, 1993, p. 25. Throughout this and the following chapter I have used the agency acronym SDIO or BMDO depending on which was in use at the time.

2. White House, Fact Sheet on Historic Economic Growth, "The Clinton Presidency: Historic Economic Growth," January 9, 2001.

3. Paul Mann, "Republicans Seek to Remake NASA," *Aviation Week & Space Technology* 141, no. 22 (December 5, 1994): 18-19.

4. James Muncy, "After the Deluge: What the GOP Takeover Could Mean for Space," fax, opinion piece written for *Space News,* published as Muncy, "After the Republican Deluge," *Space News,* December 19-25, 1994, p. 4, photocopy, file 644, *X-33* Archive (see my bibliographic essay).

5. "Clinton Space 'Blueprint' Sees Free Trade, Privatization," *Aerospace Daily,* electronic edition, September 20, 1996; James R. Asker, "Clinton Launch Policy: Upgrade ELVs, Push SSTO," *Aviation Week & Space Technology* 140, no. 17 (May 9, 1994): 24-26.

6. Tim Kyger, interview by author, tape and transcript, February 12, 1999, Universal Space Lines, Arlington, VA, p. 100, NASA Historical Reference Collection (hereafter NHRC).

7. "Kyger Materials, 1991," file 291, *X-33* Archive; Jess Sponable, comments on "X-Rocket" monograph, p. 84, file 726, *X-33* Archive; Kyger, comments on "X-Rocket" monograph, p. 84, file 753, *X-33* Archive; Chris Rosander, "SX-2: The Next Step for SSRT," September 1993, file 445, *X-33* Archive; Col. Pete Worden and Maj. Jess Sponable, "White Paper: SX-2 Advanced Technology Demonstrator (ATD) Program," January 1994, file 447, *X-33* Archive.

8. Peter V. Domenici et al. to Les Aspin, February 24, 1993, and Domenici et al. to Leon Panetta, Director, OMB, February 24, 1993, file 293, *X-33* Archive; Kyger, comments on "X-Rocket" monograph, p. 100.

9. Mark Hopkins, President, Space Cause, to Patricia Schroeder, May 17, 1993; Space Cause to Patricia Schroeder, May 17, 1993; and Space Transportation Association to Patricia Schroeder, May 12, 1993, all in file 293, *X-33* Archive.

10. Kyger, comments on "X-Rocket" monograph, p. 101.

11. Aerospace States Association, "Resolution Supporting the Funding and Continuation of the Single Stage Rocket Technology Program," ASA Resolution 93-004, file 293, *X-33* Archive.

12. Patricia Schroeder to Dana Rohrabacher, June 23, 1993, file 293, *X-33* Archive.

13. Dana Rohrabacher to Patricia Schroeder, July 2, 1993, file 293, *X-33* Archive.

14. Dana Rohrabacher, Robert Dornan, et al. to Joseph M. McDade, Subcommittee

on Defense, House Committee on Appropriations, September 7, 1993, file 293, *X-33* Archive.

15. Sen. John Warner to W. Paul Blasse, January 4, 1994, file 293, *X-33* Archive.

16. "SX-2 Advanced Technology Demonstrator SOL HQ0006-94-R-0001," September 1, 1993, file 445, *X-33* Archive.

17. Tim Kyger "Meeting Notes—the 'Dawson Meeting'," n.d.; Kyger and Allen Sherzer, "One Small Step for a Space Activist. . .," vol. 4, no. 10. October 1993, labeled "internet column," both in file 370, *X-33* Archive.

18. Ben Iannotta, "Rapid Fire DC-X Test Stalls, Funds Dry Up," *Space News*, November 1-7, 1993, p. 4.

19. Ben Iannotta, "House Proposes Shifting Some Space Programs to ARPA," *Space News*, October 11-17, 1993, p. 17.

20. Robert Dornan, Dana Rohrabacher, Norman Mineta, and Clarence Brown to John M. Deutch, October 21, 1993, file 293, *X-33* Archive.

21. Lt. Gen. Malcolm R. O'Neill to Dana Rohrabacher, November 24, 1993, file 293, *X-33* Archive; Ben Iannotta, "Committee Funds Three U.S. Launcher Options," *Space News*, November 8-14, 1993, p. 4.

22. Ben Iannotta, "U.S. Lawmakers Nix Spacelifter in Favor of ARPA Launch Vehicle," *Space News*, November 15-28, 1993, p. 4; Andrew Lawler, "U.S. Defense Department Halts Work on DC-X," *Space News*, November 1-7, 1993, p. 4.

23. Bob Walker to Al Gore, January 21, 1994; and Robert Dornan to John Deutch, January 14, 1994, both in file 293, *X-33* Archive.

24. G. Harry Stine, *Halfway to Anywhere: Achieving America's Destiny in Space* (New York: M. Evans and Co., 1996), p. 144.

25. "Pentagon Guts Funds Slated for New Launcher," *Space News*, January 10-16, 1994, p. 2. Dr. John Hamre became Deputy Secretary of Defense in July 1997. White House, press release, "Statement by the President," January 10, 2000.

26. Jess Sponable, interview by author, tape and transcript, January 16, 1998, NASA Headquarters, Washington, DC, pp. 47-49, NHRC.

27. William Boyer, "Delta Clipper Model Clears Major Hurdle," *Space News*, August 23-29, 1993, p. 3; idem, "Second Single-Stage Rocket Test May Help Draw Funds," *Space News*, September 20-26, 1993, p. 17. For an eyewitness description of the August–September 1993 test flights, see Stine, *Halfway to Anywhere*, pp. 3, 117-120, 131.

28. Boyer, "Second Single-Stage Rocket," p. 17.

29. Space Frontier Foundation, press release, January 26, 1994, file 447, *X-33* Archive.

30. Ralph M. Hall, "Opening Statement," in U.S. House of Representatives, 102d Cong., 2d sess., *1993 NASA Authorization: Hearings before the Subcommittee on Space of the Committee on Science, Space, and Technology* 2, no. 137 (Washington, DC: Government Printing Office, 1992): 3.

31. Arnold Aldrich, interview by author, tape and transcript, August 14, 1997, Lockheed Martin Corporate Headquarters, Bethesda, MD, p. 48, NHRC.

32. U.S. House of Representatives, *Conference Report*, 102d Cong., 2d sess., Report 102-902 (Washington, DC: Government Printing Office, 1992), pp. 69-70.

33. James R. Asker, "Goldin Orders Sweeping Review of NASA Programs, Eyes 30% Cuts," *Aviation Week & Space Technology* 136, no. 23 (June 1, 1992): 4-5, photocopy, file 4512, NHRC.

34. NASA, Press Release 92-154, "Goldin Announces Initiatives to Improve NASA Performance," September 17, 1992, file 4512, NHRC.

35. Aldrich, interview, p. 11.

36. Arnold D. Aldrich and Michael D. Griffin to Daniel Goldin, "Implementation Plan for 'Access to Space' Review," memorandum, January 11, 1993, file 197, *X-33* Archive. According to Ivan Bekey, the study initially was to compare shuttle upgrades and a new expendable, or partially reusable, launcher. These alternatives ultimately became Option 1 and Option 2. Bekey, interview by author, tape and transcript, March 2, 1999, NASA Headquarters, Washington, DC, p. 40, NHRC.

37. The following descriptions of the three options are from NASA Headquarters, Office of Space Systems Development, "Access to Space Study Summary Report," January 1994, pp. 8-59, file 100, *X-33* Archive.

38. Bekey, interview, p. 40.

39. "Access to Space Study Summary Report," January 1994, pp. 2-5, 8-58; Access to Space Study Advanced Technology Team, *Final Report,* vol. 1, "Executive Summary," July 1993, pp. iii, 38, both in file 85, *X-33* Archive.

40. Robert "Gene" Austin, interview by author, tape and transcript, October 6, 1997, Boeing Hangar 407, Palmdale, CA, pp. 4-6, 8-11, NHRC; Michael Griffin, interview by author, tape and transcript, August 18, 1997, Orbital Sciences Corporation, Dulles, VA, pp. 30, 33, 34, NHRC.

41. Access to Space Study Advanced Technology Team, *Final Report,* 1:I; "Access to Space Study Summary Report," p. 4.

42. Aldrich, interview, p. 55.

43. Bekey, interview, p. 41.

44. Access to Space Study Advanced Technology Team, *Final Report,* 4 vols., July 1993, files 85-88, *X-33* Archive.

45. Access to Space Study Advanced Technology Team, *Final Report,* 1:2.

46. Gary Payton, comments on "X-Rocket" monograph, February 1, 1999, n.p., file 725, *X-33* Archive.

47. Access to Space Study Advanced Technology Team, *Final Report,* vol. 4, "Operations Plan," July 1993, pp. 3-8, file 88, *X-33* Archive.

48. Austin, interview, p. 15.

49. Marshall Space Flight Center, "Experimental SSTO X Vehicle: Meeting with Industry Representatives," August 31, 1993, file 122, *X-33* Archive.

50. Austin, interview, pp. 20-21.

51. Ben Iannotta, "Winged X-2000 Project Considered: Pentagon, NASA Study Single-Stage Development Program," *Space News,* November 15-28, 1993, p. 14.

52. Kyger, interview, p. 101.

53. Sponable, comments on "X-Rocket" monograph, pp. 93, 94, 97.

54. Jess Sponable to Dr. Weiss, "NASA Access to Space Briefing," memorandum, November 3, 1993, file 372, *X-33* Archive.

55. John Cunningham, "Sabotaging SSRT's Future," *Space News,* October 18-24, 1993, p. 23; "Single Stage to Orbit Advanced Technology Demonstrator (X-2000)," August 1993, file 122, *X-33* Archive; "Single Stage to Orbit: Advanced Technology Demonstrator: SSTO Concept Proposal, X-2000," August 1993, file 162, *X-33* Archive.

TEN: The *Clipper Graham*

1. Access to Space Advanced Technology Team (Option 3), *Final Report,* vol.1, "Executive Summary," July 1993, p. 49, file 125, *X-33* Archive (see my bibliographic essay); NASA Headquarters, Office of Space Systems Development, "Access to Space Study Summary Report," January 1994, p. 71, file 100, *X-33* Archive.

2. "Access to Space Study Summary Report," p. 71.

3. Ibid., p. 69.

4. Access to Space Advanced Technology Team (Option 3), *Final Report,* 1:49.

5. Ibid.; "Access to Space Study Summary Report," p. 70.

6. NASA Headquarters, Office of Space Systems Development, OMB briefing charts, "Access to Space," October 6, 1993, file 197, *X-33* Archive.

7. See Charles Lurio, "Why We Need an SX-2 Program in DoD *Now,*" September 1994, file 449, *X-33* Archive, for an example of an argument in favor of "complementary competition" between NASA and Defense programs.

8. Director, Strategic and Space Systems, Defense Department, "Space Launch Systems Bottom-up Review," May 4, 1993, file 233, *X-33* Archive; "Air Force Presentation for the Space Launch Systems Review," April 16, 1993, file 234, *X-33* Archive.

9. Ben Iannotta, "Committee Funds Three U.S. Launcher Options," *Space News,* November 8-14, 1993, p. 4.

10.Gen. Merrill A. McPeak to Deputy Secretary of Defense and the Under Secretary of Defense for Acquisition, "Joint USAF-NASA Technology Development," memorandum, July 27, 1993, file 278, *X-33* Archive.

11. Ben Iannotta, "Winged X-2000 Project Considered: Pentagon, NASA Study Single-Stage Development Program," *Space News,* November 15-28, 1993, p. 14.

12. The NASA statement that a single-stage-to-orbit rocket would not be ready by 2005 is interestingly prescient, even though the agency had not yet started its *X-33* program whose goal was a single-stage-to-orbit prototype before the end of the decade.

13. Jess Sponable to Dr. Weiss, "NASA Access to Space Briefing," memorandum, November 3, 1993, file 372, *X-33* Archive.

14. Gary L. Denman, Director, ARPA, to Sen. Charles S. Robb, March 14, 1994, file 440, *X-33* Archive; David N. Spires, *Beyond Horizons: A Half Century of Air Force Space Leadership* (Peterson AFB: Air Force Space Command, 1997), p. 276.

15. John M. Deutch to Defense Comptroller, "FY 1994 ARPA Space Launch Funds," memorandum, January 31, 1994, file 440, *X-33* Archive; Ben Iannotta, "Backers Fight DC-X Cut," *Space News,* January 31–February 6, 1994, p. 3.

16. "Space Launch Modernization Plan: Executive Summary," April 1994, file 142, *X-33* Archive; Ben Iannotta, "Reusable vs. Expendable," *Space News,* May 16-22, 1994, p. 3.

17. Arnold Aldrich, interview by author, tape and transcript, August 14, 1997, Lockheed Martin Corporate Headquarters, Bethesda, MD, p. 56, NASA Historical Reference Collection (hereafter NHRC).

18. Ben Iannotta, "Draft Plan Defers New U.S. Rocket, *Space News,* April 4-10, 1994, p. 1.

19. Iannotta, "Winged X-2000 Project Considered," p. 14.

20. "Space Launch Modernization Plan: Executive Summary," p. 29; Ben Iannotta,

"Congress, NASA Dueling Over Reusable Rocket Management," *Space News,* May 23-29, 1994, p. 25.

21. Karen Pearce to "All Republican Members," "Space Subcommittee: September 20, 1994, National Space Transportation Policy," memorandum, October 7, 1994, file 450, *X-33* Archive; "NASA Implementation Plan for the National Space Transportation Policy," revised, November 7, 1994, file 153, *X-33* Archive. The National Space Transportation Policy also made the Departments of Commerce and Transportation responsible for fostering the international competitiveness of the U.S. launch industry.

22. Dana Rohrabacher to Al Gore, May 17, 1994, file 440, *X-33* Archive.

23. Ben Iannotta, "Liftoff Blast, Budget Limbo Hamper DC-X Program," *Space News,* July 4-10, 1994, p. 4.

24. "Defense Panels Agree to Pay for SSRT Work," *Space News,* August 15-28, 1994, p. 2.

25. Ben Iannotta, "NASA Cautious About New Rocket," *Space News,* September 26–October 2, 1994, p. 21.

26. William J. Perry to Robert Byrd, November 25, 1994, file 440, *X-33* Archive.

27. "Minutes of Senior Staff and Center Directors' Meeting," November 29, 1993, file 197, *X-33* Archive.

28. Thomas J. Lee, "Biographical Sketch," n.d., file 350, *X-33* Archive.

29. Thomas J. Lee to Daniel Goldin, "Single Stage To Orbit (SSTO) Program Planning," memorandum, undated, file 277, *X-33* Archive.

30. Ibid.

31. "NASA SSTO (R) Technology Development / Demonstration Program," October 19, 1993, briefing to Defense Department SSTO Review Team; unlabeled overview of same from Charles Camarda; Langley, "Reusable Cryo Tankage and Structures," all in file 83, *X-33* Archive.

32. Lee to Goldin.

33. Leonard David, "NASA Seeks Industry's Advice on Single-Stage Technologies," *Space News,* September 13-19, 1993, p. 12.

34. "NASA SSTO (R) Technology Development / Demonstration Program," overview 10/19/93 RMRYAN, file 83, *X-33* Archive; Jeffrey M. Lenorovitz, "Tripropellant Engine Tested for SSTO Role," *Aviation Week & Space Technology* 141, no. 2 (July 11, 1994): 54.

35. "NASA SSTO (R) Technology Development / Demonstration Program," overviews Powell 44 and Powell 51, file 83, *X-33* Archive.

36. "NASA SSTO (R) Technology Development / Demonstration Program," Austin briefing, file 83, *X-33* Archive.

37. "National Space Transportation Strategy," draft, April 26, 1994, file 153, *X-33* Archive.

38. Daniel Goldin to John Gibbons, September 22, 1993, file 278, *X-33* Archive.

39. Advanced Launch Technology Program, "SSTO Technology Program: Coordinated NASA / DoD Plan," January 21, 1994, file 164, *X-33* Archive.

40. Bob Walker to Al Gore, January 21, 1994 and Robert Dornan to John Deutch, January 14, 1994, both in file 293, *X-33* Archive; John Deutch, "FY 1994 ARPA Space Launch Funds," memorandum, January 31, 1994, file 440, *X-33* Archive.

41. George E. Brown, Jr., to Les Aspin, January 31, 1994, file 440, *X-33* Archive; Ben Iannotta, "DC-X Hangs by Thin Thread Despite Short-term Reprieve," *Space News,*

February 7-13, 1994, p. 4; Warren E. Leary, "Rocket Program Faces Budget Ax," *New York Times,* January 31, 1994, p. 13A.

42. Col. William J. Dalecky to Dana Rohrabacher, May 2, 1994, file 440, *X-33* Archive; Ben Iannotta, "Pentagon Frees Funds for More DC-X Flights," *Space News,* May 9-15, 1994, p. 4.

43. Gaubatz, interview by author, tape and transcript, October 25, 1997, Huntington Beach, CA, p. 38, NHRC.

44. DC-X Evaluation Team, "DC-X as a Demonstrator for SSTO Technologies," briefing to NASA Administrator, March 1, 1994, p. 3, file 154, *X-33* Archive.

45. Ibid., pp. 9, 17, 31.

46. The White House, Presidential Decision Directive/NSTC-4, August 5, 1994, p. 6, file 150, *X-33* Archive. The directive established "national policy, guidelines, and implementing actions for the conduct of National [*sic*] space transportation programs." (p. 1) For an informed discussion and evaluation of this policy, see U.S. Congress, Office of Technology Assessment, *The National Space Transportation Policy: Issues for Congress,* OTA-ISS-620 (Washington, DC: Government Printing Office, May 1995), pp. 13-16, 81-87.

47. "NASA Implementation Plan," p. 12.

48. NASA, Office of Inspector General, "Review of National Aeronautics and Space Administration Cooperative Agreements with Large Commercial Firms," P&A-97-001, August 22, 1997, Appendix I, p. ii, file 437, *X-33* Archive (hereafter, P&A-97-001).

49. "National Space Transportation Strategy," draft.

50. P&A-97-001, pp. 3, 5, 8, 33.

51. DC-X Evaluation Team, pp. 17, 31.

52. P&A-97-001, pp. 4-5.

53. DC-X Evaluation Team, p. 33.

54. Paul Klevatt, comments on Andrew J. Butrica, "The Spaceship That Came in from the Cold War" (unpublished manuscript), October 18, 2000, p. 3, file 858, *X-33* Archive; Ben Iannotta, "DC-X Again Takes to the Air; Funding Debate Continues," *Space News,* June 27–July 3, 1994, p. 11; "Cause of Tear in the DC-X Aeroshell Determined," *News From McDonnell Douglas,* press release 94-190, July 19, 1994, file 266, *X-33* Archive.

55. William Boyer, "Report Proposes Fates for DC-X," *Space News,* July 11-17, 1994, p. 11.

56. Paul Klevatt, "DC-X Progress/Status Report," July 5, 1994, pages "89-3/Klevatt/de/4" and "89-7/Klevatt/de/4," file 366, *X-33* Archive.

57. "Defining the X in DC-X," *Military Space* 11, no. 19 (September 19, 1994): 3, photocopy, file 276, *X-33* Archive; Liz Tucci, "Budget Shifting Reflects NASA's Changing Priorities," *Space News,* June 13-19, 1994, p. 22.

58. Jan C. Monk, Marshall Space Flight Center, "NRA 8-11 Industry Briefing," February 18, 1994, n.p., file 126, *X-33* Archive.

59. Janet Wilhelm, Marshall Space Flight Center, "NRA 8-12 Industry Briefing," February 18, 1994, n.p., file 126, *X-33* Archive; Charles Camarda, Marshall Space Flight Center, "NRA 8-12 Industry Briefing," February 18, 1994, n.p., file 126, *X-33* Archive; and S. R. Riccitiello, Ames Research Center, "NRA 8-12 Industry Briefing," February 18, 1994, n.p., file 126, *X-33* Archive.

60. Advanced Launch Technology, "Industry Briefing," February 18, 1994, file 126, *X-33* Archive.

61. Rodger Romans and Mark Stiles, Marshall Space Flight Center, "NRA 8-11 & 12 Industry Briefing," February 18, 1994, n.p., file 126, *X-33* Archive.

62. "Reusable Launch Vehicle Technology Program: Vehicle Technologies Status Report, Fiscal Year 1994," p. 1, file 22, *X-33* Archive.

63. "Defining the X in DC-X," p. 3.

64. Delma C. Freeman, Jr., Theodore A. Talay, and R. Eugene Austin, "Reusable Launch Vehicle Technology Program," IAF 96-V.4.01, p. 3, paper read at the 47th International Astronautical Congress, October 7-11, 1996, Beijing, China, photocopy, file 92, *X-33* Archive.

65. Ibid.

66. Dan Dumbacher, interview by author, tape and transcript, October 6, 1997, Boeing Hangar 407, Palmdale, CA, pp. 3-15, NHRC.

67. This description of the *Clipper Graham* incident is from "Executive Summary," in "NASA DC-XA Clipper Graham Mishap Investigation Report," September 12, 1996, p. 4-1, file 79, *X-33* Archive. A more detailed account of the damage is provided on pp. 6-2 to 6-4.

68. Ibid., p. 4-2.

69. Ibid., p. 7-9.

70. Ibid., p. 4-2.

71. Anthony Spear, *NASA Faster, Better, Cheaper Task Final Report* (Washington, DC: NASA, 2000). See, also, Arthur G. Stephenson, *Report on Project Management in NASA* (Washington, DC: NASA, 2000); NASA, "Administrator Praises Work of Review Teams," Press Release 00-37, March 13, 2000; and Howard E. McCurdy, *Faster, Better, Cheaper: Low-Cost Innovation in the U.S. Space Program* (Baltimore: Johns Hopkins University Press, 2001).

72. Jess Sponable, comments on Butrica, "The Spaceship That Came in from the Cold War," December 2000, p. 496, file 856, *X-33* Archive.

73. For example, Steve Hoeser, interview by author, tape and transcript, November 2, 1999, Boeing, Alexandria, VA, p. 18, NHRC.

74. In August 1996, Jim Muncy, then on Rep. Rohrabacher's staff, believed that funding could be pieced together from various programs to build not another *DC-XA*, but the *DC-XB* or the *DC-XC*. Each variation would have tested technologies that would contribute eventually to the construction of a full-scale prototype single-stage-to-orbit transport. Jim Muncy to Daniel Goldin, via Gary Payton, "Out of DC-XA's Ashes . . . an X-38?" memorandum, August 2, 1996, file 457, *X-33* Archive.

Conclusion

1. Robert "Gene" Austin, interview by author, tape and transcript, October 6, 1997, Boeing Hangar 407, Palmdale, CA, pp. 47-48, 50-51, NASA Historical Reference Collection (hereafter NHRC).

2. Gary Payton to Ed Frankle, NASA General Counsel, "Selection for X-33 Phase II: Design and Demonstration Cooperative Agreement Notice 8-3," fax, July 3, 1996, file 630, *X-33* Archive.

3. David Urie, interview by author, tape and transcript, September 23, 1999, Anderson, CA, pp. 9, 18, 22-23, 26, NHRC.

4. Lockheed Martin attempted to garner government funding for the ill-fated project by turning to its traditional customer, the Defense Department. "NASA Considering X-33 Engine Tests," *Space News*, June 25, 2001, pp. 1, 20.

5. Chris W. Johnson to author, e-mail, January 11, 2002.

6. NASA Facts, "The Space Launch Initiative: Technology to Pioneer the Space Frontier," June 2001.

7. Marc Selinger, "Air Force, NASA Studying Joint Development of New Reusable Launch Vehicles," *Aerospace Daily*, January 25, 2002, article 197714 [electronic edition].

8. "BMDO's Name Changed to Missile Defense Agency," *Aerospace Daily*, January 7, 2002, article 196406 [electronic edition].

9. I use the terms "micropolitical" and "macropolitical" in a manner analogous to the terms "microeconomic" and "macroeconomic."

10. Eva C. Freedman, ed., *MIT Lincoln Laboratory: Technology in the National Interest* (Cambridge: MIT Lincoln Laboratory, 1995), pp. 273-276, provides some of the basic information presented here on laboratory organization. Interviews with various laboratory staff, including a former director, conducted by the author for Lincoln Laboratory were an additional valuable source of information.

11. William Sims Bainbridge, *The Spaceflight Revolution: A Sociological Study* (New York: John Wiley & Sons, 1976), p. 198. I am grateful to the anonymous reviewer of my manuscript for this reference.

12. Roger D. Launius and Howard E. McCurdy, "Epilogue," in Launius and McCurdy, eds., *Spaceflight and the Myth of Presidential Leadership* (Urbana: University of Illinois Press, 1997), pp. 231-234.

13. Launius and McCurdy, p. 235. Jim Muncy also made the point that while space has been a partisan issue, "space has always risen and fallen on the waves of *ideology.*" Muncy, "After the Deluge: What the GOP Takeover Could Mean for Space," *Space News*, December 19-25, 1994, p. 4.

14. Launius and McCurdy, pp. 235-238.

15. Ibid., p. 238.

16. Ibid., p. 240.

Bibliographic Essay

Finding primary records on the SDIO's SSTO program and the *Delta Clipper* in traditional repositories proved to be largely unrewarding. During a visit to the McDonnell Douglas facility in Huntington Beach, California, I asked about the possibility of seeing company documents. The company official refused my request, but did provide me with a videotape and a few other items. He also showed me a hallway lined with boxes. Boeing recently had bought McDonnell Douglas and they were about to relocate. Shortly thereafter, the key members of the *DC-X* team either retired or took positions with other space businesses. The fate of those records is still unclear.

The Strategic Defense Initiative Organization (now the Missile Defense Agency) logically should have had project records as well. But Donald Baucom, the agency's historian, assured me that he did not have them. Nonetheless, he was able to supply interview transcripts relevant to the creation of the Strategic Defense Initiative and general materials, such as press releases, the McDonnell Douglas press kit, and a BMDO report to Congress, as well as key photographs.

Eventually I came to meet two individuals who had been essential to the *DC-X* project. Neither had worked for either McDonnell Douglas or the Strategic Defense Initiative Organization. Max Hunter, who lived in California, did not travel often to Washington, D.C., so Steve Hoeser, a former air force officer then working in the capital for General Research Corporation, a consulting firm, became Hunter's spokesperson in Washington. He also was (Ret.) Lt. Gen. Daniel O. Graham's assistant in transforming the *SSX* concept into a funded government program. Hoeser kept meticulous notes on meetings and briefings and collected a wealth of documentation on single-stage-to-orbit, reusable, and other rocket systems.

Tim Kyger was crucial in the effort to keep the *DC-X* program funded as a member of Rep. Dana Rohrabacher's staff. After working as a Senate staff member for two years, he joined Universal Space Lines, then under former astronaut and *DC-X* flight manager Charles "Pete" Conrad. Because he was at the center of congressional efforts to keep the *DC-X* program alive, Kyger's enormous personal collection of documents was a priceless source of program documentation.

Both Hoeser and Kyger eagerly and graciously furnished generous portions of their documentary caches dealing with the *DC-X*. A third individual, Doug Stanley, contributed vital technical documents. Stanley was a researcher at NASA's Langley Research Center detailed to the *X-33* project at NASA Headquarters. He had had a hand in the Langley review of the SDIO SSTO program Phase I industry designs. Stanley made available the Langley *DC-X* records, which constituted a rich technical documentation of Phase I industry efforts.

These three donations formed the indispensable core of primary source material on the *DC-X* and the SDIO SSTO program. They also formed a considerable share of the materials organized as part of the *X-33* history project into the *X-33* Archive. That collection included such documents relevant to the *DC-X* story as: the Access to Space, *X-2000*, National Space Transportation Policy, Moorman report, and SSTO technology program papers donated by Cindy Bruno and Matt Crouch (both NASA Headquarters); records on the space movement from Trudy Bell (writer and editor); Phase I of the SDIO SSTO program, *DC-X* flight tests, and *DC-XA* structures and cryogenic tanks documents contributed by Gary Payton (NASA Headquarters); papers on single-stage-to-orbit concepts from 1972 to 1995 provided by Theodore A. Talay (Langley Research Center); articles on the McDonnell Douglas Rapid Prototyping Department from Bill Gaubatz (Universal Space Lines) and Paul Klevatt (McDonnell Douglas, retired); copies of manuscripts and publications furnished by Max Hunter; and documents from the Ronald Reagan Presidential Library donated by W. D. "Woody" Kay (Northeastern University). Needless to say, the archive also holds documents on the *X-33* program.

The reader thus will find throughout this work's endnotes numerous references to records in the *X-33* Archive. The archive includes an electronic guide to the files and their contents on CD. Upon the termination of the *X-33* history project, I turned over the guide and archive to the NASA History Office, which in turn deposited the materials with the Washington National Records Center, National Archives and Records Administration, Suitland, Maryland. The relevant accession number is 255-01-0645. Researchers may consult the *X-33* files in the same manner as they would request any other NASA records at Suitland. The electronic guide does not indicate in which box one might find a given file; however, that information is available through Jane Odom, NASA Chief Archivist, in the NASA History Office. Researchers also should note that the History Office retained some materials, such as oral history interview transcripts, in its own Historical Reference Collection (cited in the endnotes as NHRC).

Secondary historical works on the *DC-X* do not exist. This is the first publication to treat the subject from that perspective. Nonetheless, readers might be interested in Jess Sponable, Dan Dumbacher, and Dale Shell, "The Legacy of Clipper Graham," *Aerospace America* 35 (October 1997): 26-31. Sponable was a program manager for the SDIO; Dumbacher was the NASA *DC-XA* project manager. The science fiction writer and aerospace engineer G. Harry Stine was an involved observer and vocal advocate of the SDIO SSTO program. He recorded his observations and feelings in *Halfway to Anywhere: Achieving America's Destiny in Space* (New York: M. Evans and Co., 1996), which includes a bibliography on pages 278-287. Critical comments on Stine's book by Jess Sponable and Gary Payton are in the *X-33* Archive. A piece of fiction that the

DC-X inspired also might interest the reader. The author, Daniel Graham, Jr., is the son of Lt. Gen. Daniel O. Graham. *The Gatekeepers* (Riverdale, NY: Baen Publishing; Distributed by Simon & Schuster, 1996), with an introduction by former astronaut Buzz Aldrin, attempts to show readers what would have been possible if the full-scale *DC-X* and the Brilliant Pebbles Strategic Defense Initiative architecture had been built.

Index